Lecture Notes in Computer Science 11631

Commenced Publication in 1973
Founding and Former Series Editors:
Gerhard Goos, Juris Hartmanis, and Jan van Leeuwen

More information about this series at http://www.springer.com/series/7407

Konstantin Avrachenkov ·
Paweł Prałat · Nan Ye (Eds.)

Algorithms and Models for the Web Graph

16th International Workshop, WAW 2019
Brisbane, QLD, Australia, July 6–7, 2019
Proceedings

 Springer

Editors
Konstantin Avrachenkov
Inria
Sophia Antipolis, France

Paweł Prałat
Ryerson University
Toronto, ON, Canada

Nan Ye
The University of Queensland
Brisbane, QLD, Australia

ISSN 0302-9743   ISSN 1611-3349   (electronic)
Lecture Notes in Computer Science
ISBN 978-3-030-25069-0   ISBN 978-3-030-25070-6   (eBook)
https://doi.org/10.1007/978-3-030-25070-6

LNCS Sublibrary: SL1 – Theoretical Computer Science and General Issues

This Springer imprint is published by the registered company Springer Nature Switzerland AG
The registered company address is: Gewerbestrasse 11, 6330 Cham, Switzerland

Preface

The 16th Workshop on Algorithms and Models for the Web Graph (WAW 2019) took place at the University of Queensland, Brisbane, Australia, July 6–7, 2019. This is an annual meeting, which is traditionally co-located with another, related, conference. WAW 2019 was co-located with the 20th INFORMS Applied Probability Society Conference. Co-location of the two workshops provided opportunities for researchers in two different but interrelated areas to interact and to exchange research ideas. It was an effective venue for the dissemination of new results and for fostering research collaborations.

The World Wide Web has become part of our everyday life, and information retrieval and data mining on the Web are now of enormous practical interest. The algorithms supporting these activities combine the view of the Web as a text repository and as a graph, induced in various ways by links among pages, hosts, and users. The aim of the workshop was to further the understanding of graphs that arise from the Web and various user activities on the Web, and stimulate the development of high-performance algorithms and applications that exploit these graphs. The workshop gathered the researchers who are working on graph-theoretic and algorithmic aspects of related complex networks, including social networks, citation networks, biological networks, molecular networks, and other networks arising from the Internet.

This volume contains the papers presented during the workshop. Each submission was reviewed by Program Committee members. Papers were submitted and reviewed using the EasyChair online system. The committee members decided to accept nine papers.

July 2019

Konstantin Avrachenkov
Paweł Prałat
Nan Ye

Organization

General Chairs

Andrei Z. Broder   Google Research, USA
Fan Chung Graham   University of California San Diego, USA

Organizing Committee

Konstantin Avrachenkov   Sophia Antipolis Inria, France
Paweł Prałat   Ryerson University, Canada
Nan Ye   The University of Queensland, Australia

Sponsoring Institutions

The University of Queensland
ACEMS (Australian Research Council Center of Excellence for Mathematical
and Statistical Frontiers)
Sophia Antipolis Inria
Google
Moscow Institute of Physics and Technology
Internet Mathematics

Program Committee

Konstantin Avratchenkov   Inria, France
Paolo Boldi   University of Milan, Italy
Anthony Bonato   Ryerson University, Canada
Milan Bradonjic   Bell Laboratories, USA
Fan Chung Graham   UC San Diego, USA
Collin Cooper   King's College London, UK
Andrzej Dudek   Western Michigan University, USA
Alan Frieze   Carnegie Mellon University, USA
David Gleich   Purdue University, USA
Jeannette Janssen   Dalhousie University, Canada
Bogumil Kaminski   Warsaw School of Economics, Poland
Ravi Kumar   Google, USA
Marc Lelarge   Inria, France
Stefano Leonardi   Sapienza University of Rome, Italy
Nelly Litvak   University of Twente, The Netherlands
Michael Mahoney   UC Berkeley, USA
Oliver Mason   NUI Maynooth, Ireland

Dieter Mitsche	University of Nice Sophia-Antipolis, France
Peter Morters	University of Bath, UK
Tobias Mueller	Utrecht University, The Netherlands
Liudmila Ostroumova	Yandex, Russia
Pan Peng	TU Dortmund, Germany
Xavier Perez-Gimenez	University of Nebraska-Lincoln, USA
Pawel Pralat	Ryerson University, Canada
Vittorio Scarano	University of Salerno, Italy
Przemyslaw Szufel	SGH Warsaw School of Economics, Poland
Yana Volkovich	AppNexus, USA
Nan Ye	The University of Queensland, Australia
Stephen Young	Pacific Northwest National Laboratory, USA

Contents

Using Synthetic Networks for Parameter Tuning in Community Detection

Liudmila Prokhorenkova[1,2]([✉])

[1] Moscow Institute of Physics and Technology, Dolgoprudny, Russia
[2] Yandex, Moscow, Russia
ostroumova-la@yandex.ru

Abstract. Community detection is one of the most important and challenging problems in network analysis. However, real-world networks may have very different structural properties and communities of various nature. As a result, it is hard (or even impossible) to develop one algorithm suitable for all datasets. A standard machine learning tool is to consider a parametric algorithm and choose its parameters based on the dataset at hand. However, this approach is not applicable to community detection since usually no labeled data is available for such parameter tuning. In this paper, we propose a simple and effective procedure allowing to tune hyperparameters of any given community detection algorithm without requiring any labeled data. The core idea is to generate a synthetic network with properties similar to a given real-world one, but with known communities. It turns out that tuning parameters on such synthetic graph also improves the quality for a given real-world network. To illustrate the effectiveness of the proposed algorithm, we show significant improvements obtained for several well-known parametric community detection algorithms on a variety of synthetic and real-world datasets.

Keywords: Community detection · Parameter tuning ·
Hyperparameters · LFR benchmark

1 Introduction

Community structure, which is one of the most important properties of complex networks, is characterized by the presence of groups of vertices (called communities or clusters) that are better connected to each other than to the rest of the network. In social networks, communities are formed based on common interests or on geographical location; on the Web, pages are clustered based on their topics; in protein-protein interaction networks, clusters are formed by proteins having the same specific function within the cell, and so on. Being able to identify communities is important for many applications: recommendations in social networks, graph compression, graph visualization, etc.

The problem of community detection has several peculiarities making it hard to formalize and, consequently, hard to develop a good solution for. First, as

© Springer Nature Switzerland AG 2019
K. Avrachenkov et al. (Eds.): WAW 2019, LNCS 11631, pp. 1–15, 2019.
https://doi.org/10.1007/978-3-030-25070-6_1

pointed out in several papers, there is no universal definition of communities [9]. As a result, there are no standard procedures for comparing the performance of different algorithms. Second, real-world networks may have very different structural properties and communities of various nature. Hence, it is impossible to develop one algorithm suitable for all datasets, as discussed in, e.g., [23]. A standard machine learning tool applied in such cases is to consider a parametric algorithm and tune its parameters based on the given dataset. Parameters which have to be chosen by the user based on the observed data are usually called *hyperparameters* and are often tuned via cross-validation, but this procedure requires a training part of the datasets with available ground truth labels. However, the problem of community detection is unsupervised, i.e., no ground truth community assignments are given, so standard tuning approaches are not applicable and community detection algorithms are often non-parametric.

We present a surprisingly simple and effective method for tuning hyperparameters of any community detection algorithm which requires no labeled data and chooses suitable parameters based only on the structural properties of a given graph. The core idea is to generate a synthetic network with properties similar to a given real-world one, but with known community assignments, hence we can optimize the hyperparameters on this synthetic graph and then apply the obtained algorithm to the original real-world network. It turns out that such a trick significantly improves the performance of the initial algorithm.

To demonstrate the effectiveness and the generalization ability of the proposed approach, we applied it to three different algorithms on various synthetic and real-world networks. In all cases, we obtained substantial improvements compared to the algorithms with default parameters. However, since communities in real-world networks cannot be formally defined, it is impossible to provide any theoretical guarantees for those parameter tuning strategies which do not use labeled data. As a result, the quality of any parameter tuning algorithm can be demonstrated only empirically. Based on the excellent empirical results obtained, we believe that the proposed approach captures some intrinsic properties of real-world communities and would generalize to other datasets and algorithms.

2   Background and Related Work

During the past few years, many community detection algorithms have been proposed, see [6,7,9,17] for an overview. In this section, we take a closer look at the algorithms and concepts used in the current research.

2.1   Modularity

Let us start with some notation. We are given a graph $G = (V, E)$, V is a set of n vertices, E is a set of m undirected edges. Denote by $\mathcal{C}$ a partition of V into several disjoint communities: $\mathcal{C} = \{C_1, \ldots, C_k\}$. Also, let m_{in} and m_{out} be the number of intra- and inter-cluster edges in a graph G partitioned according $\mathcal{C}$. Finally, $d(i)$ denotes the degree of a vertex i and $D(C) = \sum_{i \in C} d(i)$ is the overall degree of a community $C \in \mathcal{C}$.

Modularity is a widely used measure optimized by many community detection algorithms. It was first proposed in [21] and is defined as follows

$$Q(\mathcal{C}, G, \gamma) = \frac{m_{in}}{m} - \frac{\gamma}{4m^2} \sum_{C \in \mathcal{C}} D(C)^2, \tag{1}$$

where γ is a resolution parameter [13]. The intuition behind modularity is the following: the first term in (1) is the fraction of intra-cluster edges, which is expected to be relatively high for good partitions, while the second term penalizes this value for having too large communities. Namely, the value $\frac{\sum_{C \in \mathcal{C}} D(C)^2}{4m^2}$ is the expected fraction of intra-cluster edges if we preserve the degree sequence but connect all vertices randomly, i.e., if we assume that our graph is constructed according to the configuration model [19].

Modularity was originally introduced with $\gamma = 1$ and many community detection algorithms maximizing this measure were proposed. However, it was shown in [8] that modularity has a resolution limit, i.e., algorithms based on modularity maximization are unable to detect communities smaller than some size. Adding a resolution parameter allows to overcome this problem: larger values of γ in general lead to smaller communities. However, tuning γ is a challenging task. In this paper, we propose a solution to this problem.

2.2 Modularity Optimization and Louvain Algorithm

Many community detection algorithms are based on modularity optimization. In this paper, as one of our base algorithms, we choose arguably the most well-known and widely used greedy algorithm called Louvain [4]. It starts with each vertex forming its own community and works in several phases. To create the first level of a partition, we iterate through all vertices and for each vertex v we compute the gain in modularity coming from removing v from its community and putting it to each of its neighboring communities; then we move v to the community with the largest gain, if it is positive. When we cannot improve modularity by such local moves, the first level is formed. After that, we replace the obtained communities by supervertices connected by weighted edges; the weight between two supervertices is equal to the number of edges between the vertices of the corresponding communities. Then we repeat the process described above with the supervertices and form the second level of a partition. After that, we merge the supervertices again, and so on, as long as modularity increases. The Louvain algorithm is quite popular since it is fast and was shown to provide partitions of good quality. However, by default, it optimizes modularity with $\gamma = 1$, therefore, it suffers from a resolution limit.

2.3 Likelihood Optimization Methods

Likelihood optimization algorithms are also widely used in community detection. Such methods are mathematically sound and have theoretical guarantees under some model assumptions [3]. The main idea is to assume some underlying random graph model parameterized by community assignments and find a partition $\mathcal{C}$ that maximizes the likelihood $P(G|\mathcal{C})$, which is the probability that a graph generated according to the model with communities $\mathcal{C}$ exactly equals G.

The standard random graph model assumed by likelihood maximization methods is the stochastic block model (SBM) or its simplified version—planted partition model (PPM). In these models, the probability that two vertices are connected by an edge depends only on their community assignments. Recently, the degree-corrected stochastic block model (DCSBM) together with the degree-corrected planted partition model (DCPPM) were proposed [12]. These models take into account the observed degree sequence of a graph, and, as a result, they are more realistic. It was also noticed that if we fix the parameters of DCPPM, then likelihood maximization based on this model is equivalent to modularity optimization with some γ [22]. Finally, in a recent paper [24] the independent LFR model (ILFR) was proposed and analyzed. It was shown that ILFR gives a better fit for a variety of real-world networks [24]. In this paper, to illustrate the generalization ability of the proposed hyperparameter tuning strategy, in addition to the Louvain algorithm, we also use parametric likelihood maximization methods based on PPM and ILFR.

2.4   LFR Model

Our parameter tuning strategy is based on constructing a synthetic graph structurally similar to the observed network. To do this, we use the LFR model [14] which is the well-known synthetic benchmark usually used for comparison of community detection algorithms. LFR generates a graph with power-law distributions of both degrees and community sizes in the following way. First, we generate the degrees of vertices by sampling them independently from the power-law distribution with exponent γ_d, mean $\bar{d}$ and with maximum degree d_{max}. Then, using a mixing parameter $\hat{\mu}$, $0 < \hat{\mu} < 1$, we obtain internal and external degrees of vertices: we expect each vertex to share a fraction $1 - \hat{\mu}$ of its edges with the vertices of its community and a fraction $\hat{\mu}$ with the other vertices of the network. After that, the sizes of the communities are sampled from a power-law distribution with exponent γ_C and minimum and maximum community sizes C_{min} and C_{max}, respectively. Then, vertices are assigned to communities such that the internal degree of any vertex is less than the size of its community. Finally, the configuration model [19] with rewiring steps is used to construct a graph with a given degree sequence and with the required fraction of internal edges. The detailed description of this procedure can be found in [14].

3   Tuning Parameters

Assume that we are given a graph G and our aim is to find a partition $\mathcal{C}$ of its vertex set into disjoint communities. To do this, we have a community detection algorithm $\mathcal{A}_\theta$, where $\theta \in \Theta$ is a set of hyperparameters. Let θ_0 be the default hyperparameters. Assume that we are also given a quality function $Q(\mathcal{C}_{\mathcal{A}_\theta}, \mathcal{C}_{GT})$ allowing to measure goodness of a partition $\mathcal{C}_{\mathcal{A}_\theta}$ obtained by $\mathcal{A}_\theta$ compared to the ground truth partition $\mathcal{C}_{GT}$. Ideally, we would like to find $\bar{\theta} = \arg\max_\theta Q(\mathcal{C}_{\mathcal{A}_\theta}, \mathcal{C}_{GT})$. However, we cannot do this since $\mathcal{C}_{GT}$ is not available. Therefore, we propose to construct a synthetic graph G' which has structural properties similar to G and also has known community assignments. For this purpose, we use the LFR model described in Sect. 2.4. To apply this model, we have to define its parameters, which can be divided into *graph-based* $(n, \gamma_d, \bar{d}, d_{max})$ and *community-based* $(\gamma_C, C_{min}, C_{max}, \hat{\mu})$.

Graph-based parameters are easy to estimate:

- $n = |V(G)|$ is the number of vertices in the observed network;
- $\bar{d} = \frac{2|E(G)|}{n}$ is the average degree;
- d_{max} is the maximum degree in G;
- γ_d is the exponent of the power-law degree distribution; we estimate this parameter by fitting the power-law distribution to the cumulative degree distribution (we minimize the sum of the squared residuals in log-log scale).

Community-based parameters contain some information about the community structure, which is not known for the graph G. However, we can try to approximate these parameters by applying the algorithm $\mathcal{A}_{\theta_0}$ with default parameters to G. This would give us some partition $\mathcal{C}_0$ which can be used to estimate the remaining parameters:

- $\hat{\mu} = \frac{m_{out}}{m}$ is the mixing parameter, i.e., the fraction of inter-community edges in G partitioned according to $\mathcal{C}_0$;
- γ_C is the exponent of the power-law community size distribution; we estimate this parameter by fitting the power-law distribution to the cumulative community size distribution obtained from $\mathcal{C}_0$ (we minimize the sum of the squared residuals in log-log scale);
- C_{min} and C_{max} are the minimum and maximum community sizes in $\mathcal{C}_0$.

We generate a graph G' according to the LFR model with parameters specified above. Using G' we can tune the parameters to get a better value of θ:

$$\theta_{opt} = \arg\max_\theta Q(\mathcal{C}'_{\mathcal{A}_\theta}, \mathcal{C}'_{GT}), \tag{2}$$

where $\mathcal{C}'_{GT}$ is known ground truth partition for G' and $\mathcal{C}'_{\mathcal{A}_\theta}$ is a partition of G' obtained by $\mathcal{A}_\theta$. It turns out that this simple idea leads to a universal method for tuning θ, which successfully improves the results of several algorithms $\mathcal{A}_\theta$ on a variety of synthetic and real-world datasets, as we show in Sect. 4.

The detailed description of the proposed procedure is given in Algorithm 1. Note that in addition to the general idea described above we also propose two modifications improving the robustness of the algorithm. The first one reduces the effect of randomness in the LFR benchmark: if the number of vertices in G is small, then a network generated by the LFR model can be noisy and the optimal parameters θ_{opt} computed according to Eq. (2) may vary from sample to

---

Algorithm 1. Hyperparameter tuning

---

input :   graph G, algorithm $\mathcal{A}_\theta$, default hyperparameters θ_0, candidate
                parameters $\{\theta_i\}_{i=1}^l$, quality function Q, n_{graphs}, n_{runs}

$n, \bar{d}, d_{max}, \gamma_d \leftarrow EstimateGraphParams(G)$;
$\mathcal{C}_0 \leftarrow \mathcal{A}_{\theta_0}(G)$;
$\hat{\mu}, \gamma_C, C_{min}, C_{max} \leftarrow EstimateCommunityParams(G, \mathcal{C}_0)$;
$ParamsList \leftarrow \emptyset$;
for $i \leftarrow 1$ **to** n_{graphs} **do**
    $G', \mathcal{C}'_{GT} \leftarrow GenerateLFR(n, \bar{d}, d_{max}, \gamma_d, \hat{\mu}, \gamma_C, C_{min}, C_{max})$;
    $QualityList \leftarrow \emptyset$;
    for $\theta \in \{\theta_i\}_{i=1}^l$ **do**
        $Qualities \leftarrow \emptyset$;
        for $j \leftarrow 1$ **to** n_{runs} **do**
            $\mathcal{C}_\theta \leftarrow \mathcal{A}_\theta(G')$;
            Add $Q(\mathcal{C}_\theta, \mathcal{C}'_{GT})$ to $Qualities$;
        $MeanQuality \leftarrow mean(Qualities)$;
        Add $MeanQuality$ to $QualityList$;
    $index \leftarrow \arg\max(QualityList)$;
    Add θ_{index} to $ParamsList$;
$\theta = median(ParamsList)$;
return θ

---

sample. Hence, we propose to generate n_{graphs} synthetic networks and take the median of the obtained parameters. The value n_{graphs} depends on computational resources: larger values, obviously, lead to more stable results. Fortunately, as we discuss in Sect. 4.5, this effect of randomness is critical only for small graphs, so we do not have to increase computational complexity much for large datasets.

The second improvement accounts for a possible randomness of the algorithm $\mathcal{A}_\theta$. If $\mathcal{A}_\theta$ includes some random steps, then we can increase the robustness of our procedure by running $\mathcal{A}_\theta$ several times and averaging the obtained qualities. The corresponding parameter is called n_{runs} in Algorithm 1. Formally, in this case Eq. (2) should be replaced by

$$\theta_{opt} = \arg\max_\theta \frac{1}{n_{runs}} \sum_{i=1}^{n_{runs}} Q(\mathcal{C}'_{\mathcal{A}_\theta, i}, \mathcal{C}'_{GT}), \tag{3}$$

where $\mathcal{C}'_{\mathcal{A}_\theta, i}$ is a (random) partition obtained by $\mathcal{A}_\theta$ on G'. If $\mathcal{A}_\theta$ is deterministic, then it is sufficient to take $n_{runs} = 1$.

Note that for the sake of simplicity in Algorithm 1 we use grid search to approximately find θ_{opt} defined in (3). However, any other method of black-box optimization can be used instead, e.g., random search [2], Bayesian optimization [25], Gaussian processes [10], sequential model-based optimization [11], and so on. More advanced black-box optimization methods can significantly speed up the algorithm.

Let us discuss the time complexity of the proposed algorithm. If complexity of $\mathcal{A}_\theta$ is $f(G)$, then complexity of Algorithm 1 is $O\left(f(G) \cdot l \cdot n_{runs} \cdot n_{graphs}\right)$, where l is the number of steps made by the black-box optimization (the complexity of generating G' is usually negligible compared with community detection). In other words, the complexity is $n_{runs} \cdot n_{graphs}$ times larger than the complexity of any black-box parameter optimization algorithm. However, as we discuss in Sect. 4.5, n_{runs} and n_{graphs} can be equal to one for large datasets.

Finally, note that it can be tempting to make several iterations of Algorithm 1 to further improve θ_{opt}. Namely, in Algorithm 1 we estimate community-based parameters of LFR using the partition $\mathcal{C}_0$ obtained with $\mathcal{A}_{\theta_0}$. Then, we obtain better parameters θ_{opt}. These parameters can be further used to get a better partition using $\mathcal{A}_{\theta_{opt}}$ and this partition is expected to give even better community-based parameters. However, in our preliminary experiments, we did not notice significant improvements from using several iterations, therefore we propose to use Algorithm 1 as it is without increasing its computational complexity.

4   Experiments

4.1   Parametric Algorithms

We use the following algorithms to illustrate the effectiveness of the proposed hyperparameter tuning strategy.

Louvain. This algorithm is described in Sect. 2.2, it has the resolution parameter γ with default value $\gamma_0 = 1$. We take the publicly available implementation from [24],[1] where the algorithm is called DCPPM since modularity maximization is equivalent to the likelihood optimization for the DCPPM random graph model.

PPM. This algorithms is based on likelihood optimization for PPM (see Sect. 2.3). We use the publicly available implementation proposed in [24], where the Louvain algorithm is used as a basis to optimize the likelihood for several models. Since likelihood optimization for PPM is equivalent to maximizing a simplified version of modularity based on the Erdős–Rényi model instead of the configuration model [22], PPM algorithm also has a resolution parameter γ with the default value $\gamma_0 = 1$.

ILFR. This is a likelihood optimization algorithm based on the ILFR model (see Sect. 2.3). Again, we use the publicly available implementation from [24]. ILFR algorithm has one parameter μ called mixing parameter and no default value for this parameter is proposed in the literature. In this paper, we take $\mu_0 = 0.3$, which is close to the average mixing parameter in the real-world datasets under consideration (see Sect. 4.2). Our experiments confirm that $\mu_0 = 0.3$ is a reasonable default value for this algorithm.

Let us stress that in this paper we are not aiming to develop the best community detection algorithm or to analyze all existing methods. Our main goal is to

---

[1] https://github.com/altsoph/community_loglike.

Table 1. Real-world datasets

Dataset	n	m	Num. clusters	Mixing parameter
Karate club [27]	34	78	2	0.128
Dolphin network [16]	62	159	2	0.038
College football [21]	115	613	11	0.325
Political books [20]	105	441	3	0.159
Political blogs [1]	1224	16715	2	0.094
email-Eu-core [15]	986	16064	42	0.664
Cora citation [26]	24166	89157	70	0.458
AS [5]	23752	58416	176	0.561

show that hyperparameter tuning is possible in the field of community detection. We use several base algorithms described above to illustrate the generalization ability of the proposed approach. For each algorithm, our aim is to improve its default parameter by our parameter tuning strategy.

4.2 Datasets

Synthetic Networks. We generated several synthetic graphs according to the LFR benchmark described in Sect. 2.4 with $n = 10^4$, $\gamma_d = 2.5$, $\bar{d} = 20$, $d_{max} = 200$, $\gamma_C = 1.5$, $C_{min} = 50$, $C_{max} = 500$, $\hat{\mu} \in \{0.4, 0.5, 0.6, 0.7\}$.[2] On the one hand, one would expect results obtained on such synthetic datasets to be optimistic, since the same LFR model is used both to tune the parameters and to validate the performance of the algorithms. On the other hand, recall that the most important ingredient of the model, i.e., the distribution of community sizes, is not known and has to be estimated using the initial community detection algorithm, and incorrect estimates may negatively affect the final performance.

Real-World Networks. We follow the work [24], where the authors collected and shared 8 real-world datasets publicly available in different sources.[3] For all these datasets, the ground truth community assignments are available and the communities are non-overlapping. These networks are of various sizes and structural properties, see the description in Table 1.

4.3 Evaluation Metrics

In the literature, there is no universally accepted metric for evaluating the performance of community detection algorithms. Therefore, we analyze several standard ones [7]. Namely, we use two widely used similarity measures based on

---

[2] Note that $\hat{\mu} > 0.5$ does not mean the absence of community structure since usually a community is much smaller than the rest of the network and even if more than a half of the edges for each vertex go outside the community, the density of edges inside the community is still large.

[3] https://github.com/altsoph/community_loglike/tree/master/datasets.

counting correctly and incorrectly classified pairs of vertices: Rand and Jaccard indices. We also consider the Normalized Mutual Information (NMI) of two partitions: if NMI is close to 1, one needs a small amount of information to infer the ground truth partition from the obtained one, i.e., two partitions are similar.

4.4   Experimental Setup

We apply the proposed strategy to the algorithms described in Sect. 4.1. We use the grid search to find the parameter θ_{opt} (we do this to make our results easier to reproduce and we also need this for the analysis of stability in Sect. 4.5). For ILFR we try μ in the range $[0, 1]$ with step size 0.05 and for Louvain and PPM on real-world datasets we take γ in the range $[0, 2]$ with step size 0.1. Although we noticed that in some cases the optimal γ for PPM and Louvain can be larger than 2, such cases rarely occur on real-world datasets. On synthetic graphs, we take γ in the range $[0, 4]$ (with step size 0.2) to demonstrate the behavior of γ_{opt} depending on $\hat{\mu}$.

To guarantee stability and reproducibility of the obtained results, we choose a sufficiently large parameter n_{runs}, although we noticed similar improvements with much smaller values. Namely, for Karate, Dolphins, Football, and Political books we take $n_{runs} = 10^3$, for Political blogs and Eu-core $n_{runs} = 100$, for Cora, AS, and synthetic networks $n_{runs} = 2$. We take $n_{graphs} = 10^3$ for four smallest datasets and $n_{graphs} = 100$ for the other ones (we choose such large values to plot the histograms on Fig. 1).

Finally, note that it is impossible to measure the statistical significance of obtained improvements on real-world datasets since we have only one copy for each graph. However, we can account for the randomness included in the algorithms. Namely, Louvain, PPM, and ILFR are randomized, since at each iteration they order the vertices randomly. Therefore, to measure if θ_{opt} is significantly better or worse than θ_0, we can run each algorithm several times and then apply the unpaired t-test (we use 100 runs in all cases).

4.5   Results

In this section, we first discuss the improvements obtained for each algorithm and then analyze the stability of the parameter tuning strategy and the effect of the parameter n_{graphs}.

Louvain Algorithm. In Table 2, for each similarity measure we present the value for the baseline algorithm (with $\gamma = 1$), the value for the tuned algorithm, and the obtained parameter γ_{opt}. Since Louvain is randomized, we provide the mean value together with an estimate of the standard deviation, which is given in brackets. The number of runs used to compute these values depends on the size of the dataset and on the available computational resources: 10^4 for Karate, Dolphins, Football and Political books, 10^3 for Political blogs and Eu-core, 100 for Cora, AS and synthetic datasets.

Table 2. Louvain algorithm, default value is $\gamma_0 = 1$, standard deviation is given in the brackets

	Rand			Jaccard			NMI		
Dataset	Default	Tuned	γ_{opt}	Default	Tuned	γ_{opt}	Default	Tuned	γ_{opt}
Karate	0.76 (0.02)	**0.95** (0.02)	0.6	0.52 (0.04)	**0.89** (0.03)	0.5	0.63 (0.05)	**0.74** (0.07)	0.7
Dolphins	0.65 (0.02)	**0.87** (0.07)	0.5	0.37 (0.04)	**0.61** (0.13)	0.1	0.52 (0.04)	0.52 (0.04)	1.0
Football	0.97 (0.01)	**0.99** (0.00)	1.7	0.72 (0.06)	**0.90** (0.04)	1.7	0.92 (0.02)	**0.97** (0.01)	1.7
Pol. books	0.83 (0.02)	**0.85** (0.00)	0.8	0.61 (0.06)	**0.65** (0.01)	0.8	0.54 (0.02)	**0.56** (0.01)	0.8
Pol. blogs	0.88 (0.00)	**0.90** (0.00)	0.7	0.78 (0.01)	**0.82** (0.00)	0.7	0.64 (0.01)	**0.68** (0.01)	0.8
Eu-core	0.86 (0.02)	**0.93** (0.00)	1.4	0.22 (0.02)	**0.35** (0.01)	1.4	0.58 (0.02)	**0.66** (0.01)	1.4
Cora	0.94 (0.00)	**0.96** (0.00)	2.0	0.13 (0.00)	**0.15** (0.00)	2.0	0.46 (0.01)	**0.49** (0.00)	2.0
AS	0.82 (0.00)	**0.82** (0.00)	1.8	0.19 (0.03)	**0.26** (0.01)	0.6	0.49 (0.01)	0.49 (0.01)	0.8
LFR-0.4	1.00 (0.00)	**1.00** (0.00)	2.8	0.96 (0.04)	**1.00** (0.00)	2.8	0.99 (0.00)	**1.00** (0.00)	2.8
LFR-0.5	1.00 (0.00)	**1.00** (0.00)	3.0	0.86 (0.08)	**1.00** (0.01)	3.0	0.98 (0.01)	**1.00** (0.00)	3.0
LFR-0.6	0.98 (0.01)	**1.00** (0.00)	3.6	0.61 (0.12)	**0.97** (0.01)	3.6	0.94 (0.02)	**0.99** (0.00)	3.6
LFR-0.7	0.91 (0.01)	**0.98** (0.00)	3.8	0.09 (0.02)	**0.32** (0.03)	3.6	0.39 (0.05)	**0.68** (0.02)	3.8

One can see that our tuning strategy improves (or does not change) the results in all cases and the obtained improvements can be huge. For example, on Karate we obtain remarkable improvements from 0.761 to 0.945 (relative change is 24%) according to Rand and from 0.52 to 0.892 (72%) according to Jaccard; on Dolphins we get 35% improvement for Rand and 63% for Jaccard; on Football we obtain plus 25% for Jaccard; and so on. As discussed in Sect. 4.4, we measured the statistical significance of the obtained improvements. The results which are significantly better are marked in bold in Table 2. On real-world datasets all improvements are statistically significant (p-value $\ll 0.01$).[4] Let us note that in many cases the results of the tuned algorithm are much better than the best results reported in [24], where the authors used other strategies for choosing the hyperparameter values.[5]

For synthetic datasets, we also observe huge improvements and all of them are statistically significant. While for $\hat{\mu} \in \{0.4, 0.5\}$ the default algorithm can be considered as good enough, for large values of $\hat{\mu}$, $\hat{\mu} \in \{0.6, 0.7\}$, the tuned one is much better. For example, for LFR-0.7 the tuned parameter gives Jaccard index almost 4 times larger than the default one.

We noticed that for most of the datasets the values of γ_{opt} computed using different similarity measures are the same or close to each other. However, there are some exceptions. The first one is Dolphins, where for Jaccard $\gamma_{opt} = 0.1$, for Rand $\gamma_{opt} = 0.5$, for NMI $\gamma_{opt} = 1.0$. We checked that if we take the median value $\gamma_{opt} = 0.5$, then for all measures we obtain statistically significant improvements,

---

[4] The results in Tables 2, 3 and 4 are rounded to two decimals, so there may be a statistically significant improvement even when the numbers in the table are equal. Also, standard deviation less than 0.005 is rounded to zero.

[5] For small datasets, our results for the default Louvain algorithm may differ from the ones reported in [24]. The reason is the high values of standard deviation. The authors of [24] averaged the results over 5 runs of the algorithm, while we use more runs, i.e., our average values are more stable.

Table 3. PPM algorithm, default value $\gamma_0 = 1$, standard deviation is given in the brackets

Dataset	Rand			Jaccard			NMI		
	Default	Tuned	γ_{opt}	Default	Tuned	γ_{opt}	Default	Tuned	γ_{opt}
Karate	0.76 (0.02)	**0.78** (0.04)	0.8	**0.51** (0.04)	0.49 (0.00)	0.1	0.63 (0.05)	0.63 (0.05)	1.0
Dolphins	0.62 (0.03)	**0.76** (0.04)	0.7	0.33 (0.04)	**0.82** (0.19)	0.1	**0.47** (0.04)	0.41 (0.02)	1.6
Football	0.97 (0.01)	**0.99** (0.00)	1.6	0.72 (0.04)	**0.90** (0.04)	1.6	0.92 (0.01)	**0.97** (0.01)	1.6
Pol. books	0.78 (0.02)	**0.85** (0.01)	0.7	0.48 (0.04)	**0.65** (0.02)	0.7	0.50 (0.02)	**0.57** (0.01)	0.7
Pol. blogs	0.65 (0.02)	**0.72** (0.04)	0.4	0.32 (0.02)	**0.47** (0.04)	0.4	0.29 (0.02)	**0.33** (0.04)	0.5
Eu-core	**0.80** (0.02)	0.77 (0.02)	0.9	**0.10** (0.01)	0.09 (0.01)	0.9	**0.53** (0.02)	0.49 (0.02)	0.8
Cora	0.94 (0.00)	**0.96** (0.00)	2.0	0.11 (0.00)	**0.13** (0.00)	2.0	0.47 (0.00)	**0.50** (0.00)	2.0
AS	0.79 (0.01)	**0.81** (0.00)	1.8	0.11 (0.01)	**0.15** (0.03)	0.8	0.46 (0.02)	0.46 (0.02)	1.2
LFR-0.4	1.00 (0.00)	**1.00** (0.00)	2.8	0.99 (0.02)	**1.00** (0.02)	2.8	1.00 (0.00)	**1.00** (0.00)	2.8
LFR-0.5	1.00 (0.01)	**1.00** (0.00)	3.0	0.88 (0.13)	**0.96** (0.05)	3.0	0.99 (0.01)	**0.99** (0.01)	3.0
LFR-0.6	0.97 (0.02)	**0.99** (0.00)	3.2	0.44 (0.20)	**0.67** (0.09)	3.2	0.85 (0.11)	**0.91** (0.04)	3.0
LFR-0.7	0.80 (0.03)	**0.97** (0.01)	2.8	0.03 (0.02)	**0.15** (0.07)	2.8	0.18 (0.11)	**0.51** (0.11)	2.8

which seems to be another way to increase the stability of our strategy. The most notable case, where γ_{opt} significantly differs for different similarity measures, is AS dataset, where $\gamma_{opt} = 1.8 > \gamma_0$ for Rand, $\gamma_{opt} = 0.6 < \gamma_0$ for Jaccard, and $\gamma_{opt} = 0.8 < \gamma_0$ for NMI. We will further make similar observations for other algorithms on this dataset. Such instability may mean that this dataset does not have a clear community structure (which can sometimes be the case for real-world networks [18]).

PPM Algorithm. For PPM (Table 3), our strategy improves the original algorithm for all real-world datasets but Eu-core (for all similarity measures), Karate (only for Jaccard), and Dolphins (only for NMI). Note that Karate and Dolphins are the only datasets (except for AS, which will be discussed further in this section), where the obtained values for γ_{opt} are quite different for different similarity measures. We checked that if for these two datasets we take the median value of γ_{opt}, (0.8 for Karate and 0.7 for Dolphins), then we obtain improvements in all six cases, five of them, except NMI on Karate, are statistically significant (p-value $\ll 0.01$). On Eu-core the quality of PPM with $\gamma_0 = 1$ is worse than the quality of Louvain with $\gamma = 1$. This seems to be the reason why PPM chooses a suboptimal parameter γ_{opt}: a partition obtained by PPM does not allow for a good estimate of the community-based parameters. As for Louvain, in many cases the obtained improvements are huge: e.g., the relative improvement for the Jaccard index is 147% on Dolphins, 26% on Football, 35% on Political books, 50% on Political blogs, an so on. All improvements are statistically significant.

We also improve the default algorithm on all synthetic datasets and for all similarity measures. As for the Louvain algorithm, the improvements are especially huge for large $\hat{\mu}$, $\hat{\mu} \in \{0.6, 0.7\}$. All improvements are statistically significant.

ILFR Algorithm. For real-world datasets, in almost all cases, we obtain significant improvements (see Table 4). One exception is Dolphins for NMI. This, again, can be fixed by taking a median of the values μ_{opt} obtained for all sim-

Table 4. ILFR algorithm, default value $\mu_0 = 0.3$, standard deviation is given in the brackets

Dataset	Rand			Jaccard			NMI		
	Default	Tuned	μ_{opt}	Default	Tuned	μ_{opt}	Default	Tuned	μ_{opt}
Karate	0.75 (0.03)	**0.85** (0.04)	0.15	0.51 (0.04)	**0.74** (0.07)	0.05	0.63 (0.06)	0.63 (0.06)	0.30
Dolphins	0.58 (0.01)	**0.62** (0.03)	0.15	0.25 (0.02)	**0.56** (0.00)	0.00	**0.45** (0.02)	0.26 (0.00)	1.00
Football	0.99 (0.00)	0.99 (0.00)	0.45	0.91 (0.03)	0.91 (0.02)	0.45	0.97 (0.01)	0.97 (0.00)	0.45
Pol. books	0.73 (0.02)	**0.82** (0.01)	0.15	0.35 (0.04)	**0.59** (0.03)	0.15	0.45 (0.01)	**0.53** (0.02)	0.15
Pol. blogs	0.77 (0.02)	**0.85** (0.04)	0.15	0.57 (0.05)	**0.73** (0.04)	0.15	0.44 (0.01)	**0.53** (0.03)	0.20
Eu-core	0.89 (0.02)	**0.94** (0.01)	0.50	0.23 (0.03)	**0.37** (0.02)	0.50	0.64 (0.02)	**0.71** (0.01)	0.50
Cora	**0.98** (0.00)	0.98 (0.00)	0.05	0.06 (0.00)	**0.10** (0.00)	0.05	**0.55** (0.00)	0.43 (0.01)	0.00
AS	**0.83** (0.00)	0.83 (0.00)	1.00	0.02 (0.00)	**0.18** (0.00)	0.00	**0.44** (0.00)	0.42 (0.00)	1.00
LFR-0.4	1.00 (0.00)	1.00 (0.00)	0.40	1.00 (0.00)	1.00 (0.00)	0.40	1.00 (0.00)	1.00 (0.00)	0.40
LFR-0.5	1.00 (0.00)	1.00 (0.00)	0.35	1.00 (0.01)	1.00 (0.01)	0.35	1.00 (0.00)	1.00 (0.00)	0.35
LFR-0.6	1.00 (0.00)	1.00 (0.00)	0.25	0.96 (0.08)	0.97 (0.06)	0.25	0.99 (0.01)	**1.00** (0.00)	0.25
LFR-0.7	0.97 (0.02)	**0.98** (0.01)	0.35	0.35 (0.13)	0.34 (0.12)	0.30	0.74 (0.06)	0.74 (0.06)	0.30

ilarity measures on this dataset: $\mu_{opt} = 0.15$ improves the results compared to $\mu_0 = 0.3$ for all three measures. Other bad examples are Cora and AS, where Rand and NMI decrease, while Jaccard increases. For all other datasets, we obtain improvements. In many cases, the difference is huge and statistically significant. On synthetic datasets, the default ILFR algorithm is the best among the considered ones. In some cases, however, the default algorithm is further improved by our hyperparameter tuning strategy, while in others the difference is not statistically significant. Surprisingly, for large values of $\hat{\mu}$ the tuned value μ_{opt} is much smaller than $\hat{\mu}$. For example, for $\hat{\mu} = 0.6$ we get $\mu_{opt} = 0.25$, although we checked that the estimated parameter used for generating synthetic graphs is very close to 0.6.

For real-world and synthetic networks, the obtained value μ_{opt} can be both larger and smaller than $\mu_0 = 0.3$. Also, for synthetic networks, μ_0 is close to the obtained μ_{opt}. We conclude that the chosen default value is reasonable.

In rare cases, μ_{opt} for a dataset can be quite different for different similarity measures. On AS, $\mu_{opt} = 0$ for Jaccard and $\mu_{opt} = 1$ for Rand and NMI. Note that if $\mu = 0$, then the obtained algorithm tends to group all vertices in one cluster, while for $\mu = 1$ all vertices form their own clusters. Interestingly, for the Jaccard index, such a trivial partition outperforms the default algorithm. Moreover, the algorithm putting each vertex in its own cluster has close to the best performance according to the Rand index compared to all algorithms discussed in this section (both default and tuned). We conclude that AS does not have a clear community structure.

Stability of Generated Graphs. As discussed in Sect. 3, there are two sources of possible noise in the proposed parameter tuning procedure: (1) for small graphs the generated LFR network can be noisy, which may lead to unstable predictions of θ_{opt}, (2) the randomness of $\mathcal{A}$ may also affect the estimate of θ_{opt} in Eq. (3). The effect of the second problem can be understood using Tables 2, 3 and 4, where the standard deviations for θ_0 and θ_{opt} are presented.

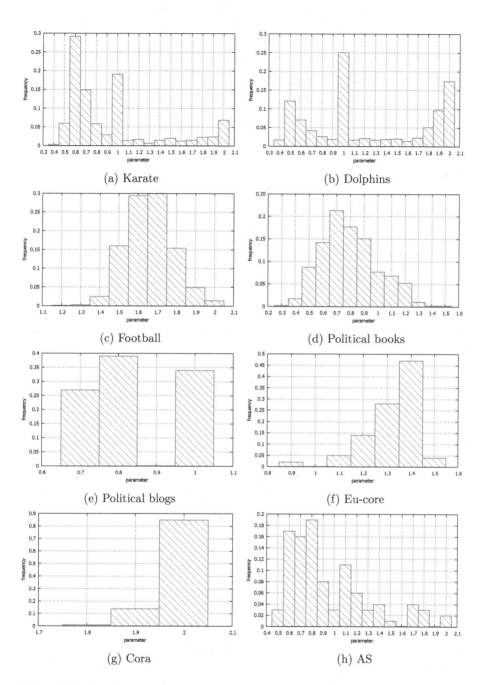

Fig. 1. The distribution of γ_{opt} for the Louvain algorithm, NMI similarity measure

To analyze the effect of noise caused by the randomness in LFR graphs and to show that it is more pronounced for small datasets, we looked at the distribution of the parameters θ_{opt} obtained for different samples of generated graphs. We demonstrate this effect using the Louvain algorithm and NMI similarity measure (see Fig. 1), we take $n_{graphs} = 10^3$ for four smallest datasets and $n_{graphs} = 100$ for the other ones. Except for the AS dataset, which is noisy according to all our experiments, one can clearly see that the variance of γ_{opt} decreases when n increases. As a result, we see that for large datasets even $n_{graphs} = 1$ already provides a good estimate for γ_{opt}.

5   Conclusion

We proposed and analyzed a surprisingly simple yet effective algorithm for hyper-parameter tuning in community detection. The core idea is to generate a synthetic graph structurally similar to the observed network but with known community assignments. Using this graph, we can apply any standard black-box optimization strategy to approximately find the optimal hyperparameters and use them to cluster the original network. We empirically demonstrated that such a trick applied to several algorithms leads to significant improvements on both synthetic and real-world datasets. Now, being able to tune parameters of any community detection algorithm, one can develop and successfully apply parametric community detection algorithms, which was not previously possible.

Acknowledgements. This study was funded by the Russian Foundation for Basic Research according to the research project 18-31-00207 and Russian President grant supporting leading scientific schools of the Russian Federation NSh-6760.2018.1.

References

1. Adamic, L.A., Glance, N.: The political blogosphere and the 2004 us election: divided they blog. In: Proceedings of the 3rd International Workshop on Link Discovery, pp. 36–43. ACM (2005)
2. Bergstra, J., Bengio, Y.: Random search for hyper-parameter optimization. J. Mach. Learn. Res. **13**(Feb), 281–305 (2012)
3. Bickel, P.J., Chen, A.: A nonparametric view of network models and newman-girvan and other modularities. Proc. Natl. Acad. Sci. **106**(50), 21068–21073 (2009)
4. Blondel, V.D., Guillaume, J.L., Lambiotte, R., Lefebvre, E.: Fast unfolding of communities in large networks. J. Stat. Mech.: Theory Exp. **2008**(10), P10008 (2008)
5. Boguná, M., Papadopoulos, F., Krioukov, D.: Sustaining the internet with hyperbolic mapping. Nat. Commun. **1**, 62 (2010)
6. Coscia, M., Giannotti, F., Pedreschi, D.: A classification for community discovery methods in complex networks. Stat. Anal. Data Min.: ASA Data Sci. J. **4**(5), 512–546 (2011)
7. Fortunato, S.: Community detection in graphs. Phys. Rep. **486**(3), 75–174 (2010)
8. Fortunato, S., Barthélemy, M.: Resolution limit in community detection. Proc. Natl. Acad. Sci. **104**(1), 36–41 (2007)

9. Fortunato, S., Hric, D.: Community detection in networks: a user guide. Phys. Rep. **659**, 1–44 (2016)
10. Golovin, D., Solnik, B., Moitra, S., Kochanski, G., Karro, J., Sculley, D.: Google vizier: a service for black-box optimization. In: International Conference on Knowledge Discovery and Data Mining, pp. 1487–1495. ACM (2017)
11. Hutter, F., Hoos, H.H., Leyton-Brown, K.: Sequential model-based optimization for general algorithm configuration. In: Coello, C.A.C. (ed.) LION 2011. LNCS, vol. 6683, pp. 507–523. Springer, Heidelberg (2011). https://doi.org/10.1007/978-3-642-25566-3_40
12. Karrer, B., Newman, M.E.: Stochastic blockmodels and community structure in networks. Phys. Rev. E **83**(1), 016107 (2011)
13. Lancichinetti, A., Fortunato, S.: Limits of modularity maximization in community detection. Phys. Rev. E **84**(6), 066122 (2011)
14. Lancichinetti, A., Fortunato, S., Radicchi, F.: Benchmark graphs for testing community detection algorithms. Phys. Rev. E **78**(4), 046110 (2008)
15. Leskovec, J., Kleinberg, J., Faloutsos, C.: Graph evolution: densification and shrinking diameters. ACM Trans. Knowl. Discov. Data (TKDD) **1**(1), 2 (2007)
16. Lusseau, D., Schneider, K., Boisseau, O.J., Haase, P., Slooten, E., Dawson, S.M.: The bottlenose dolphin community of doubtful sound features a large proportion of long-lasting associations. Behav. Ecol. Sociobiol. **54**(4), 396–405 (2003)
17. Malliaros, F.D., Vazirgiannis, M.: Clustering and community detection in directed networks: a survey. Phys. Rep. **533**(4), 95–142 (2013)
18. Miasnikof, P., Prokhorenkova, L., Shestopaloff, A.Y., Raigorodskii, A.: A statistical test of heterogeneous subgraph densities to assess clusterability. In: 13th LION Learning and Intelligent OptimizatioN Conference. Springer (2019)
19. Molloy, M., Reed, B.: A critical point for random graphs with a given degree sequence. Random Struct. Algorithms **6**(2–3), 161–180 (1995)
20. Newman, M.E.: Modularity and community structure in networks. Proc. Natl. Acad. Sci. **103**(23), 8577–8582 (2006)
21. Newman, M.E., Girvan, M.: Finding and evaluating community structure in networks. Phys. Rev. E **69**(2), 026113 (2004)
22. Newman, M.: Community detection in networks: modularity optimization and maximum likelihood are equivalent. arXiv preprint arXiv:1606.02319 (2016)
23. Peel, L., Larremore, D.B., Clauset, A.: The ground truth about metadata and community detection in networks. Sci. Adv. **3**(5), e1602548 (2017)
24. Prokhorenkova, L., Tikhonov, A.: Community detection through likelihood optimization: in search of a sound model. In: The World Wide Web Conference, pp. 1498–1508. ACM (2019)
25. Snoek, J., et al.: Scalable Bayesian optimization using deep neural networks. In: International Conference on Machine Learning, pp. 2171–2180 (2015)
26. Šubelj, L., Bajec, M.: Model of complex networks based on citation dynamics. In: Proceedings of the 22nd International Conference on World Wide Web, pp. 527–530. ACM (2013)
27. Zachary, W.W.: An information flow model for conflict and fission in small groups. J. Anthropol. Res. **33**(4), 452–473 (1977)

Efficiency of Transformations
of Proximity Measures
for Graph Clustering

Rinat Aynulin[1,2(✉)]

[1] Kotel'nikov Institute of Radio-engineering and Electronics (IRE) of Russian
Academy of Sciences, Mokhovaya 11-7, Moscow 125009, Russia
[2] Moscow Institute of Physics and Technology, 9 Inststitutskii per., Dolgoprudny,
Moscow Region 141700, Russia
rinat.aynulin@phystech.edu

Abstract. Choice of proximity measure for the nodes greatly affects the
results of graph clustering. In this paper, we consider several proximity
measures transformed with a number of functions including the logarithmic function, the power function, and a family of activation functions. Transformations are tested in experiments in which several classical datasets are clustered using the k-Means, Ward, and the spectral
method. The analysis of experimental results with statistical methods
shows that a number of transformed proximity measures outperform
their non-transformed versions. The top-performing transformed measures are the Heat measure transformed with the power function, the
Forest measure transformed with the power function, and the Forest
measure transformed with the logarithmic function.

1   Introduction

Research in such areas as bioinformatics, chemistry, social networks, Web search,
etc. often involves work with objects which are connected to each other. Graphs
are a natural way of representing structured data: nodes represent objects while
edges represent connections between them. If we need to detect groups (clusters)
of similar objects in the graph taking into account connections between them,
clustering methods are used.

Regardless of the clustering method used, for its implementation, it is introduced, either explicitly or implicitly, a distance on the set of graph nodes. Some
common metrics are the shortest path metric and the commute time metric, however, there are also more complex metrics [9, Chapter 15], for example, those
based on the Heat, Communicability, Forest, and other proximity measures. The
selected distance directly affects the quality of clustering, and the choice of the
most suitable distance is an indispensable task of the researcher.

Considering the increase in computer processing power and the current trend
for the huge growth in the volume and variety of processed data, among which
there is a lot of data that needs to be clustered, graph clustering has many

© Springer Nature Switzerland AG 2019
K. Avrachenkov et al. (Eds.): WAW 2019, LNCS 11631, pp. 16–29, 2019.
https://doi.org/10.1007/978-3-030-25070-6_2

applications [23]. Due to the importance of graph clustering, a lot of methods for improving the quality of clustering have been suggested in recent years. This can be achieved by adjusting the clustering method. For example, the k-Means method can be improved using the local search method [17]. Another way to improve the quality of clustering is to change the metric used. In this study, we explore the potential for improvement of the quality of clustering by *transforming* metrics.

Previously, it was shown that the use of logarithmic measures (measures obtained by taking the logarithm of each element of the kernel matrix) positively affects the quality of clustering [1,16]. These results are discussed in more detail in Sect. 2. In the subsequent sections, we investigate the efficiency of various metric transformations for graph clustering.

The main result is that certain transformations improve the results of clustering in comparison with the non-transformed metrics. The set of efficient transformations depends on the transformed metric and the clustering method used, however, for most metrics and algorithms, the best quality is shown by such transformations as the logarithmic function and the power function with certain exponents.

2   Related Work

The problem of finding the most suitable metrics for graph clustering already attracted the interest of researchers. In this section, we provide a brief overview of previous studies in which different metrics have been introduced and compared.

For a long time, the attention of the researchers was focused on the shortest path distance [13]. [3] provides a short survey of various graph metrics and relationships between them. The papers [10] and [11] propose the Communicability metric and explore some of its properties; [7] introduces the Forest metric and shows its relation to the Resistance metric. The Heat metric is introduced in [19]. In [2], the Walk metric is analyzed.

As for comparing the efficiency of different metrics for graph clustering, the metrics abbreviated as MCS, WGU, UGU, MMCS, and MMCSN are introduced in [24] and their efficiency in document clustering is studied.

In [25], metrics Corrected Commute Time, Free Energy, Logarithmic Forest, Randomized shortest-path, Sigmoid commute time and Shortest-path are compared in experiments with real data. The most efficient measures were Free Energy and Randomized shortest-path.

In [28], Euclidean Commute Time is compared with the standard Euclidean Distance, and Euclidean Commute Time wins.

[1] provides comparison of various similarity measures in the context of spectral clustering on the stochastic block model, and the top-performing measures were the normalized heat-type measures with the logarithmic transformation.

In [16], the logarithmic transformation is applied to different metrics, and the best in the experiments was the metric named logarithmic Communicability: the Communicability metric transformed with the logarithmic function.

In the present paper, we examine a number of measures obtained by several transformations from the known measures.

3 Preliminaries

3.1 Definitions

Let $G = (V, E)$ be an undirected graph with a non-empty node set V and a set of edges E (i.e., 2-element subsets of V). Given the adjacency matrix A and the diagonal degree matrix $D = \mathrm{diag}(A \cdot \mathbf{1})$ (where $\mathbf{1} = (1, ..., 1)^T$), the Laplacian matrix L is defined as $L = D - A$.

A *measure* on a graph G is a function $k : V(G) \times V(G) \rightarrow \mathbb{R}$ that shows proximity or similarity between nodes of G. A *kernel* on a graph is a graph similarity measure that can be represented as a symmetric positive semidefinite matrix K, or Gram matrix [1]. The conditions to be met by a proximity measure are listed in [5,6]. The distances used in this paper are defined via the corresponding kernels. In [5], a duality between proximity measures and metrics is studied.

The distance $d_{ij} = k_{ii} + k_{jj} - k_{ij} - k_{ji}$, where $K = (k_{ij})$ is the matrix corresponding to a proximity measure, satisfies the axioms of metric. In a matrix notation, the relationship between the metric and the proximity measure can be written as $\mathcal{D} = (d_{ij}) = \mathrm{diag}(K) \cdot \mathbf{1}^T + \mathbf{1} \cdot \mathrm{diag}(K)^T - K^T$.

3.2 Kernels

Through this paper, we consider the following kernels:

- Walk: $K = \sum_{n=0}^{\infty} \alpha^n A^n = (I - \alpha A)^{-1}$, $\alpha \in (0, q^{-1})$, where q is the spectral radius of the adjacency matrix of a graph [2,18]
- Communicability: $K = \sum_{n=0}^{\infty} \frac{\alpha^n A^n}{n!} = \exp(\alpha A)$, $\alpha > 0$ [10,11]
- Forest: $K = \sum_{n=0}^{\infty} \alpha^n (-L)^n = (I + \alpha L)^{-1}$, $\alpha > 0$ [4]
- Heat: $K = \sum_{n=0}^{\infty} \frac{\alpha^n (-L)^n}{n!} = \exp(-\alpha L)$, $\alpha > 0$ [19]

3.3 Transformations

By a transformation of a metric we mean the application of some mathematical function to each element of the corresponding kernel. The transformations that we use in this paper are:

- Logarithmic function (Log): $f(x) = \log(x)$
- Power function: $f(x) = x^p$; we consider various values of $p < 1$
- Hyperbolic tangent (TanH): $f(x) = \tanh(x)$
- Sigmoid: $f(x) = \sigma(x) = 1/(1 + e^{-x})$
- Arctangent (ArcTan): $f(x) = \tan^{-1}(x)$
- Softsign: $f(x) = x/(1 + |x|)$
- Inverse Square Root Unit (ISRU): $f(x) = x/(\sqrt{1 + x^2})$
- Sigmoid-weighted Linear Unit (SiLU): $f(x) = x \cdot \sigma(x)$

4   Experiments and Results

4.1   Experimental Methodology

In this section, we compare the efficiency of metrics and their transformed versions in experiments on several classical datasets with the number of nodes ranging from 34 to 999:

- Zachary: the social network of members of the University Karate Club, described by Wayne Zachary. The nodes represent the members of the club, the edges indicate friendship relationships between them. The classes in this dataset are two groups of participants, formed as a result of a conflict in the leadership of the club. The number of nodes in the graph $n = 34$, the number of links between them $m = 78$.
- Football: the graph of teams of the US University football league in 2000. The nodes represent the teams, the edges are games between them, and the classes are athletic conferences. $n = 115$, $m = 613$.
- Polbooks: the graph of American political books purchased on Amazon. Information was collected in the run-up to the 2008 elections. The nodes represent the books, an edge means the fact that two books are often purchased by buyers together, and classes are the political ideology of the books. The number of the classes $c = 3$. $n = 105$, $m = 441$.
- Newsgroups: a collection of 20000 documents taken from 20 newsgroups of Usernet. In this paper, we use 6 graphs extracted from this dataset with $c = 2$ ($n = 400$, $n = 398$, $n = 399$) and $c = 3$ ($n = 600$, $n = 598$, $n = 595$). The nodes represents the documents, and the classes are the topics of the documents. The weight of an edge between two nodes indicates the level of commonness between them [27].

For each dataset, we consider 4 proximity measures (Walk, Communicability, Forest, and Heat) transformed with each of the 10 transformations under research, including all the transformations from Sect. 3.3 (the power function is used with exponents $= 1/3$ and $= 1/2$) and the Identity Transformation (the original measure is returned). Then the transformed proximity measure is used to cluster the graph using 3 clustering methods: the k-Means method [14], the spectral clustering method [22], and the Ward method [26]. So, each dataset is clustered using $4 \cdot 10 \cdot 3 = 120$ algorithms. The algorithm here refers to the triplet: a clustering method, a proximity measure, and a transformation.

The quality of clustering is evaluated with the ARI quality index. ARI is introduced in [15], and [20] provides justifications for choosing it. According to the definition of this quality index, the larger the value, the better. If the value of ARI is 1, then it is the perfect match; 0 refers to random labeling.

Since each of the measures used depends on its parameter, the search for the optimal parameter was carried out in the experiments.

4.2   Analysis

We analyze the results of experiments in order to determine the best transformations for the proximity measures, as well as the best algorithms (triplets) overall.

In Table 1, the top 10 results after ranking of the algorithms by the average ARI over 9 datasets are presented. When ranking this way, the top algorithm that uses a non-transformed proximity measure (spectral, Walk, Identity)[1] is ranked only 33nd out of 120 and its average ARI is 0.65. For the top algorithm that uses a transformed measure (k-Means, Heat, Power function with $p = 1/2$), the average ARI is 0.782.

Table 1. The best algorithms by the average ARI on the datasets

№	Method	Measure	Transformation	Average ARI
1	k-Means	Heat	Power function, $p = 1/2$	0.782
2	k-Means	Forest	Power function, $p = 1/2$	0.781
3	k-Means	Forest	Power function, $p = 1/3$	0.780
4	Spectral	Forest	Log	0.764
5	Spectral	Heat	Log	0.764
6	Spectral	Heat	Power function, $p = 1/3$	0.755
7	Spectral	Forest	Power function, $p = 1/3$	0.755
8	k-Means	Heat	Power function, $p = 1/3$	0.745
9	Spectral	Communicability	Log	0.744
10	Spectral	Heat	Power function, $p = 1/2$	0.740

We can also rank (the ranks of tied algorithms are averaged) all the 120 algorithms by the ARI on each dataset, and then find the average value of the rank for each algorithm.

The ranking of the algorithms by the average rank shows similar results (Table 2) to the ranking by the average ARI. The leading algorithms that use transformed proximity measures still outperform algorithms that use non-transformed proximity measures. The top algorithm with a non-transformed proximity measure (spectral, Heat measure, Identity) is ranked 30th out of 120 and its average rank is 40.25 (versus the average rank 18.46 for the top algorithm with a transformed measure (k-Means, Heat, Power function with $p = 1/2$)).

So, an overall comparison of the algorithms demonstrates that the algorithms that use transformed proximity measures outperform algorithms that use the non-transformed proximity measures.

---

[1] Hereinafter, a clustering algorithm is denoted by such a triplet. The first element in a triplet is a clustering method, the second is a proximity measure, and the third is a transformation.

Table 2. The best algorithms by the average rank on the datasets

№	Method	Measure	Transformation	Average rank
1	k-Means	Heat	Power function, $p = 1/2$	18.46
2	k-Means	Forest	Power function, $p = 1/2$	20.37
3	k-Means	Forest	Power function, $p = 1/3$	21.71
4	Spectral	Heat	Log	22.25
5	k-Means	Heat	Power function, $p = 1/3$	22.96
6	Spectral	Forest	Log	26.17
7	Spectral	Communicability	Softsign	27.58
8	Spectral	Communicability	ArcTan	28.33
9	Spectral	Communicability	Log	28.91
10	Spectral	Heat	Sigmoid	30.17

Further conclusions will be based on the ranking by the average ARI.

The best clustering quality is shown by the logarithmic function and the power function with exponents $= 1/3$ and $1/2$: 16 leading algorithms use these transformations. In Table 3, one can find the best result for each of the transformations under research among all the algorithms.

Some transformations turned out to be useless and only worsen the results in comparison with non-transformed proximity measures. For example, this is the Sigmoid-weighted linear unit (SiLU) activation function. Note that this is the fastest growing function of all the transformations studied (except for the Identity Transformation).

Table 3. The best results shown by each of the transformation among all the algorithms

Transformation	Highest average ARI	Highest position	Method
Power function, $p = 1/2$	0.782	1	k-Means
Power function, $p = 1/3$	0.780	3	k-Means
Log	0.764	4	Spectral
ArcTan	0.716	17	Spectral
Softsign	0.709	22	Spectral
TanH	0.697	23	Spectral
ISRU	0.686	24	Spectral
Sigmoid	0.662	29	Spectral
Identity Transformation	0.650	33	Spectral
SiLU	0.633	43	Spectral

In Table 4, for each method and proximity measure, the best transformation is presented. As can be seen, the top transformations are the logarithmic function and the power function with exponents 1/2 and 1/3. The Identity Transformation (i.e., the non-transformed measure) is not the best for any combination of a clustering method and a proximity measure.

Table 4. The best transformation by the average ARI for each combination of the clustering methods and proximity measures

	Method	Measure	Transformation	Average ARI
1	k-Means	Heat	Power function, $p = 1/2$	0.782
2	k-Means	Forest	Power function, $p = 1/2$	0.781
4	Spectral	Forest	Log	0.764
5	Spectral	Heat	Log	0.764
9	Spectral	Communicability	Log	0.744
11	Spectral	Walk	Log	0.740
18	Ward	Forest	Log	0.715
21	Ward	Heat	Power function, $p = 1/3$	0.710
34	Ward	Walk	Power function, $p = 1/3$	0.649
39	Ward	Communicability	Power function, $p = 1/3$	0.637
51	k-Means	Walk	Power function, $p = 1/2$	0.608
58	k-Means	Communicability	Power function, $p = 1/3$	0.575

The main purpose of this paper is to find the best transformations, but from these results, we can also make some conclusions about proximity measures and clustering methods. As for proximity measures, the leaders here are the Heat and the Forest measures. Remarkably, these measures, in contrast to the Walk and the Communicability proximity measures, are based on the Laplacian matrix.

Regarding clustering methods, the best quality is shown by the k-Means method. The spectral method also demonstrates a good performance. The best algorithm based on the Ward method is ranked only 18th out of 120.

4.3   Examining the Results by Friedman and Nemenyi Tests

In the previous section, we compared 120 clustering algorithms[2] and made some conclusions about the efficiency of transformations of proximity measures based on averaging of the quality index over all datasets.

However, this approach has several limitations. Say, averaging is susceptible to outliers. Excellent performance of an algorithm on one dataset can compensate for poor performance on the other datasets and vice versa. In general, such a behavior is not desirable, since we prefer algorithms to perform well on as many datasets as possible.

---

[2] Recall that an algorithm here refers to a triplet: a clustering method, a proximity measure, and a transformation.

Although we can draw some conclusions based on averaging over all datasets, a more reliable way to compare the algorithms is desirable to be sure that transformations really improve the results of graph clustering in comparison with the non-transformed measures.

In [8], the authors study the methods for statistical estimation of the quality of graph clustering on multiple datasets. Among the tests considered are the paired testing of algorithms, the ANOVA test, the Friedman test, and others. The authors have concluded that the Friedman test in combination with the post-hoc Nemenyi test shows the best results.

Following the same lines, we conduct the Friedman test [12] and the Nemenyi test [21] and look for the best transformation for each specific clustering method and proximity measure.

The statistical tests are performed at significance level $\alpha = 0.05$.

The null hypothesis in the Friedman test is that all transformations for a given metric and a clustering method show the same quality. If this hypothesis is rejected, then we run the Nemenyi test and compare all transformations pairwise.

In Table 5, one can find the values of the Friedman statistic, as well as the p-value (3 significant digits) for all combinations of a proximity measure and a clustering method. If the p-value is less or equal to the statistical significance $\alpha = 0.05$, then the null hypothesis is rejected. It means that there is a statistically significant difference in the quality of different transformation for the clustering algorithms that use these measure and method.

Table 5. The Friedman test results

Method	Measure	F	p-value
Ward	Walk	44.878	0.000
Ward	Communicability	36.327	0.000
Ward	Heat	66.677	0.000
Ward	Forest	63.530	0.000
k-Means	Walk	49.598	0.000
k-Means	Communicability	45.719	0.000
k-Means	Forest	53.190	0.000
k-Means	Heat	48.959	0.000
Spectral	Walk	25.658	0.002
Spectral	Communicability	19.615	0.020
Spectral	Forest	26.003	0.002
Spectral	Heat	18.079	0.034

For all the measures and methods, we obtained the p-value less than the significance level. Thus, after the Friedman test based on the results of experiments on the datasets, the null hypothesis is rejected in all cases. Therefore, the quality of clustering has a significant difference for different transformations.

The next step is a post-hoc analysis in order to find the results for which the transformations are significantly different. In the Nemenyi test, two algorithms are significantly different if the difference between their average ranks is more than CD (critical difference) [8].

In Table 6, for each combination of a measure and a method, transformations are presented which significantly improve the result of clustering as compared to the non-transformed measures.

Table 6. The Nemenyi test result: transformations that are significantly better than the Identity Transformation for each method and proximity measure

Method	Measure	Transformations that are significantly better than the Identity Transformation
Ward	Walk	Power function ($p = 1/3$)
Ward	Communicability	Power function ($p = 1/3$ and $p = 1/2$)
Ward	Heat	Power function ($p = 1/3$ and $p = 1/2$), Log, Softsign
Ward	Forest	Power function ($p = 1/3$ and $p = 1/2$), Log, Softsign
k-Means	Walk	Power function ($p = 1/3$ and $p = 1/2$)
k-Means	Communicability	Power function ($p = 1/3$ and $p = 1/2$)
k-Means	Forest	Power function ($p = 1/3$ and $p = 1/2$)
k-Means	Heat	Power function ($p = 1/3$ and $p = 1/2$)
Spectral	Walk	-
Spectral	Communicability	-
Spectral	Forest	-
Spectral	Heat	-

As can be seen, there are no transformations that are significantly better than the Identity Transformation for the spectral method used with all the proximity measures. This can be explained by the fact that when using the spectral method, even the non-transformed measures showed good quality. Although a number of transformations result in some quality improvement (e.g., for the spectral method and the Heat measure, the Logarithmic transformation gives the average ARI of 0.764 versus 0.623 for the non-transformed measure), this improvement is less than the critical difference.

In [8], a method of visual representation of the differences in quality for several clustering algorithms on several datasets was proposed. It is called the Critical Difference diagram (CD-diagram). These diagrams can give more insight into the efficiency of transformations.

The CD-diagrams for the algorithms based on the Ward method are shown in Figs. 1, 2, 3 and 4. The set of CD-diagrams for the remaining combinations of proximity measures and clustering methods can be found in Figs. 5 and 6 in Appendix A.

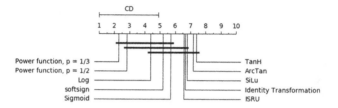

Fig. 1. The CD-diagram for the algorithms based on the Ward method and the Walk measure

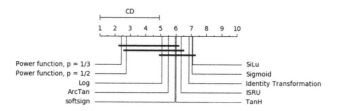

Fig. 2. The CD-diagram for the algorithms based on the Ward method and the Communicability measure

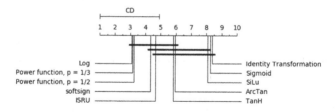

Fig. 3. The CD-diagram for the algorithms based on the Ward method and the Forest measure

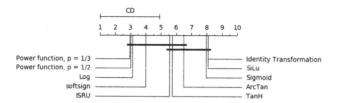

Fig. 4. The CD-diagram for the algorithms based on the Ward method and the Heat measure

The x-axis shows the values of the average rank for each transformation. The transformations that are in the left part of the diagram show the best quality in the experiments. A horizontal line connecting two or more transformations indicates that there is no statistically significant difference in the quality of clustering between them.

The results of statistical tests confirm the superiority of transformed measures over the non-transformed ones. We can identify a group of transformations that significantly improve the quality of clustering for the algorithms based on the Ward and the k-Means method. For the algorithms based on the spectral method, we cannot identify any transformations as "significantly" improving the clustering results. However, according to the previous section and the obtained CD-diagrams, the transformed measures still outperform their non-transformed versions for the spectral method.

5   Conclusion

In this paper, we studied how applying transformations to graph proximity measures such as Walk, Communicability, Heat, and Forest affects the quality of clustering. To test the efficiency of the transformations, a number of experiments were performed, where several classical datasets were clustered using the k-Means, Ward, and the spectral method. The results of the experiments were processed with the methods of non-parametric analysis of variance: the Friedman test and the Nemenyi test.

As a result, we can conclude that transformations of proximity measures significantly improve the quality of graph clustering. A set of the most efficient transformations depends on the proximity measure and the clustering method used. However, we can recommend a number of transformations that showed a good performance in all the cases: the logarithmic function and the power function with certain exponents.

We can also recommend a number of top-performing algorithms—combinations of a clustering method, a proximity measure, and a transformation—for graph vertex clustering. They are listed in Table 1.

Thus, the efficiency of a number of proximity measures that are already actively used in practice can be improved by applying simple transformations.

Appendix A

CD-diagrams

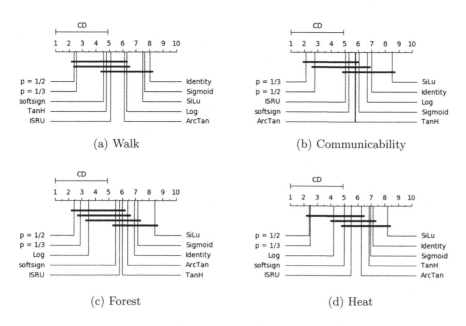

Fig. 5. The CD-diagrams for the algorithms based on the k-Means method

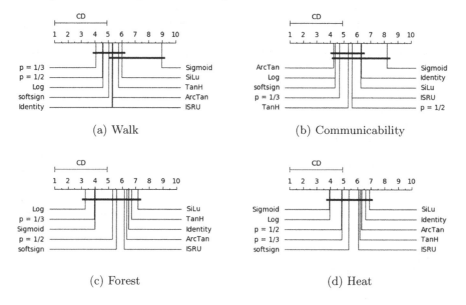

Fig. 6. The CD-diagrams for the algorithms based on the spectral method

References

1. Avrachenkov, K., Chebotarev, P., Rubanov, D.: Kernels on graphs as proximity measures. In: Bonato, A., Chung Graham, F., Prałat, P. (eds.) WAW 2017. LNCS, vol. 10519, pp. 27–41. Springer, Cham (2017). https://doi.org/10.1007/978-3-319-67810-8_3
2. Chebotarev, P.: The walk distances in graphs. Discrete Appl. Math. **160**, 1484–1500 (2012)
3. Chebotarev, P.: Studying new classes of graph metrics. In: Nielsen, F., Barbaresco, F. (eds.) GSI 2013. LNCS, vol. 8085, pp. 207–214. Springer, Heidelberg (2013). https://doi.org/10.1007/978-3-642-40020-9_21
4. Chebotarev, P., Shamis, E.: On the proximity measure for graph vertices provided by the inverse Laplacian characteristic matrix. In: Abstracts of the Conference "Linear Algebra and its Application", 10–12 June 1995, pp. 6–7 (1995)
5. Chebotarev, P., Shamis, E.: On a duality between metrics and Σ-proximities. Autom. Remote Control. **59**, 608–612 (1998)
6. Chebotarev, P., Shamis, E.: On proximity measures for graph vertices. Autom. Remote Control. **59**, 1443–1459 (1998)
7. Chebotarev, P., Shamis, E.: The forest metrics for graph vertices. Electron. Notes Discret. Math. **11**, 98–107 (2002)
8. Demšar, J.: Statistical comparisons of classifiers over multiple data sets. J. Mach. Learn. Res. **7**, 1–30 (2006)
9. Deza, M.M., Deza, E.: Encyclopedia of Distances. Springer, Berlin (2016). https://doi.org/10.1007/978-3-662-52844-0
10. Estrada, E.: The communicability distance in graphs. Linear Algebr. Its Appl. **436**, 4317–4328 (2012)
11. Fouss, F., Yen, L., Pirotte, A., Saerens, M.: An experimental investigation of graph kernels on a collaborative recommendation task. In: Proceedings of the Sixth International Conference on Data Mining (ICDM 2006), pp. 863–868 (2006)
12. Friedman, M.: The use of ranks to avoid the assumption of normality implicit in the analysis of variance. J. Am. Stat. Assoc. **32**, 675–701 (1937)
13. Goddard, W., Oellermann, O.R.: Distance in graphs. In: Dehmer, M. (ed.) Structural Analysis of Complex Networks, pp. 49–72. Birkhäuser, Boston (2010). https://doi.org/10.1007/978-0-8176-4789-6_3
14. Hartigan, J.A., Wong, M.A.: Algorithm as 136: a k-means clustering algorithm. J. R. Stat. Soc. Ser. C (Appl. Stat.) **28**(1), 100–108 (1979)
15. Hubert, L., Arabie, P.: Comparing partitions. J. Classif. **2**(1), 193–218 (1985)
16. Ivashkin, V., Chebotarev, P.: Do logarithmic proximity measures outperform plain ones in graph clustering? In: Kalyagin, V., Nikolaev, A., Pardalos, P., Prokopyev, O. (eds.) NET 2016. PROMS, vol. 197, pp. 87–105. Springer, Cham (2017). https://doi.org/10.1007/978-3-319-56829-4_8
17. Kanungo, T., Mount, D.M., Netanyahu, N.S., Piatko, C.D., Silvermank, R., Wu, A.Y.: A local search approximation algorithm for k-means clustering. Comput. Geom. **28**(2–3), 89–112 (2004)
18. Katz, L.: A new status index derived from sociometric analysis. Psychometrika **18**(1), 39–43 (1953)
19. Kondor, R.I., Lafferty, J.D.: Diffusion kernels on graphs and other discrete input spaces. In: Proceedings of ICML, pp. 315–322 (2002)
20. Milligan, G., Cooper, M.: A study of the comparability of external criteria for hierarchical cluster-analysis. Multivar. Behav. Res. **21**, 441–458 (1986)

21. Nemenyi, P.: Distribution-free multiple comparisons. Biometrics **18**(2), 263 (1962)
22. Ng, A.Y., Jordan, M.I., Weiss, Y.: On spectral clustering: analysis and an algorithm. In: Advances in Neural Information Processing Systems, pp. 849–856 (2002)
23. Schaeffer, S.E.: Graph clustering. Comput. Sci. Rev. **1**, 27–64 (2007)
24. Schenker, A., Last, M., Bunke, H., Kandel, A.: Comparison of distance measures for graph-based clustering of documents. In: Hancock, E., Vento, M. (eds.) GbRPR 2003. LNCS, vol. 2726, pp. 202–213. Springer, Heidelberg (2003). https://doi.org/10.1007/3-540-45028-9_18
25. Sommer, F., Fouss, F., Saerens, M.: Comparison of graph node distances on clustering tasks. In: Villa, A.E.P., Masulli, P., Pons Rivero, A.J. (eds.) ICANN 2016. LNCS, vol. 9886, pp. 192–201. Springer, Cham (2016). https://doi.org/10.1007/978-3-319-44778-0_23
26. Ward, J.H.: Hierarchical grouping to optimize an objective function. J. Am. Stat. Assoc. **58**, 236–244 (1963)
27. Yen, L., Fouss, F., Decaestecker, C., Francq, P., Saerens, M.: Graph nodes clustering based on the commute-time kernel. In: Zhou, Z.-H., Li, H., Yang, Q. (eds.) PAKDD 2007. LNCS (LNAI), vol. 4426, pp. 1037–1045. Springer, Heidelberg (2007). https://doi.org/10.1007/978-3-540-71701-0_117
28. Yen, L., Vanvyve, D., Wouters, F.: Clustering using a random walk based distance measure. In: Proceedings of the 13th European Symposium on Artificial Neural Networks, ESAAN-2005, pp. 317–324 (2005)

Almost Exact Recovery in Label Spreading

Konstantin Avrachenkov and Maximilien Dreveton$^{(\boxtimes)}$

Inria Sophia Antipolis, 2004 Route des Lucioles, 06902 Valbonne, France
{k.avrachenkov,maximilien.dreveton}@inria.fr

Abstract. In semi-supervised graph clustering setting, an expert provides cluster membership of few nodes. This little amount of information allows one to achieve high accuracy clustering using efficient computational procedures. Our main goal is to provide a theoretical justification why the graph-based semi-supervised learning works very well. Specifically, for the Stochastic Block Model in the moderately sparse regime, we prove that popular semi-supervised clustering methods like Label Spreading achieve asymptotically almost exact recovery as long as the fraction of labeled nodes does not go to zero and the average degree goes to infinity.

Keywords: Semi-supervised clustering · Community detection · Label spreading · Random graphs · Stochastic Block Model

1 Introduction and Previous Work

Graph clustering consists of partitioning a graph into communities (or clusters) so that nodes in the same cluster are, in some sense, more densely connected than nodes belonging to different clusters. Graph clustering (or community detection) is a fundamental problem in machine learning. Many scientific disciplines rely on graphs to model a large number of interacting agents: atoms or interacting particles in statistical physics, proteins interactions in molecular biology, social networks in sociology, the Internet's webgraph in computer science, etc. Such complex networks typically have clustering structure, whose detection and description is very important for network analysis.

To model complex networks, we can interpret them as random graphs. The simplest random graph model with clustering structure is the Stochastic Block Model (SBM), introduced independently in [6] and [9]. SBM is a generalization of the Erdős-Rényi (ER) random graph [7,8]. In its easiest form, an SBM graph has two communities of equal size, and edges between nodes of the same community are drawn with probability p, and edges between nodes of different communities have a probability q, where $p \neq q$. Of course, this is a very basic model of a graph with clustering structure. Despite its simplicity, the basic SBM poses a number of theoretical challenges for community detection problem and highlights various intuitions and trade-offs.

© Springer Nature Switzerland AG 2019
K. Avrachenkov et al. (Eds.): WAW 2019, LNCS 11631, pp. 30–43, 2019.
https://doi.org/10.1007/978-3-030-25070-6_3

Community detection in SBM is still a very active topic, and one can find a recent and complete review in [1], mentioning the up to date unsupervised clustering results. In this paper, we will consider a semi-supervised situation, where an oracle reveals the community belonging of a fraction of nodes. In practice, labeling nodes according to their community requires human intervention, thus is expensive (could be months of experiments in a case of protein study), and the fraction of pre-labeled nodes is expected to be the smallest possible. As was noted in the previous publications on graph-based semi-supervised learning (see e.g., [2,5,14–16]), it is a very powerful technique allowing to achieve high accuracy with only a small number of labeled data points. Moreover, as those methods are naturally distributed, they can efficiently cluster large graphs.

A popular graph based semi-supervised method is Label Spreading [14]. The main goal of the present work is to provide a theoretical justification why Label Spreading works well, by showing that it achieves almost exact recovery on SBM graphs, in the moderately sparse regime (when the average degree d is of the order of $\log n$), as long as the fraction of labeled points r does not go to zero.

Note that the recovery is said to be exact if all nodes are correctly labeled (almost surely, in the limit as n goes to infinity), and almost exact if the fraction of misclassified nodes goes to 0 (almost surely, when n goes to infinity) [1].

The paper is structured as follows: in Sect. 2, we describe the minimization procedure we used for semi-supervised graph clustering (Label Spreading) and provide more background references on the semi-supervised learning. In Sect. 3 we study the case of SBM graphs, using a mean field analysis. We derive the exact expression for the semi-supervised solution of the mean field SBM and explain why exact recovery is possible for the mean field. Then, we show concentration of the limit towards its mean field value and conclude with the recovery result. Section 4 provides discussion and directions for future research.

2   Semi-supervised Graph Clustering with the Normalized Laplacian Matrix (Label Spreading)

Let $G = (V, E)$ be a graph, where V is the set of n nodes, and E is the set of m edges. In the following, we will consider weighted undirected networks: each edge $(ij) \in E$ holds a positive weight w_{ij}. Thus, the graph can be fully represented by a symmetric matrix W, where the entry (ij) of W is the weight w_{ij} of an edge between nodes i and j (a weight of zero corresponds to the absence of edge). When the weights are binary, the weight matrix is called the adjacency matrix and is traditionally denoted by A. The degree d_i of a node $i \in V$ is defined as the sum of the weights of all edges going from i, that is $d_i = \sum_j w_{ij}$. The diagonal matrix D with entries d_i is called the degree matrix.

We will consider a graph exhibiting a community structure: hence, the set of nodes can be partitioned into K non overlapping communities (or clusters). By observing only V and E, and supposing K known, we aim to recover the underlying partition in a semi-supervised manner. This means some nodes are already labeled: we know to which community they belong. Let ℓ and u be respectively the set of labeled node and the set of unlabeled nodes. Without loss of generality, we can suppose that the first $|\ell|$ nodes are labeled, and we define r the ratio of labeled nodes with respect to the total number of nodes ($|\ell| = r|V|$).

Our strategy is to find a matrix X of size $n \times K$ from which we could predict the node's labels. We will refer to the columns $X_{.k}$ as classification functions, and node i will be classified in cluster $k(i)$ if:

$$k(i) = \arg\max_{k' \in \{1,...,K\}} X_{ik'}. \tag{1}$$

To make use of the semi-supervised setting, we shall fix the values of X on the labeled data. More precisely, we introduce the $n \times K$ ground-truth matrix Y as:

$$Y_{ik} = \begin{cases} 1 & \text{if node } i \text{ is in community } k \\ 0 & \text{otherwise.} \end{cases}$$

Since $Y_{\ell.}$ is known, where $Y_{\ell.}$ denotes the first $|\ell|$ rows of the matrix Y (corresponding to the labeled nodes), we will enforce $X_{\ell.} = Y_{\ell.}$. The other rows of X, denoted $X_{u.}$, will be chosen to minimize the energy function:

$$E(X) := \mathrm{tr}\left(X^T \mathcal{L} X\right) \tag{2}$$
$$\text{such that } X_{\ell.} = Y_{\ell.} \tag{3}$$

where $\mathcal{L} := I_n - D^{-\frac{1}{2}} W D^{-\frac{1}{2}}$ is the normalized Laplacian of the graph.

The choice to minimize an energy function to solve a semi-supervised learning problem can be traced back to [16]. In that paper, the authors chose a standard Laplacian-based energy function. In later works (see e.g., [2,11,14]) it has been shown that one can achieve a better accuracy with the use of the normalized Laplacian. There is another important argument why we have chosen to focus on the normalized Laplacian method: as it will be clear from the ensuing development, the normalized Laplacian's spectral norm concentrates sufficiently well around its expectation [12].

The minimization problem (2)–(3) can be solved using Lagrange multiplier:

$$L(X) := E(X) + \lambda \, \mathrm{tr}\left((X_{\ell.} - Y_{\ell.})^T (X_{\ell.} - Y_{\ell.})\right). \tag{4}$$

To compute the solution explicitly in a matrix form, we split the weight matrix W (and other matrices like D) into four blocks $\begin{pmatrix} W_{\ell\ell} & W_{\ell u} \\ W_{u\ell} & W_{uu} \end{pmatrix}$, where $W_{\ell\ell}$ is a sub-matrix corresponding to the first $|\ell|$ rows and columns of matrix W. The solution $X = \begin{pmatrix} X_\ell \\ X_u \end{pmatrix}$ of the optimization problem (2)–(3) can be derived by letting the partial derivatives of the convex function L with respect to X_{ik} (for $i \notin \ell$ and $k \in \{1, \ldots, K\}$) being zero, and writing the solution in a matrix form. More precisely, let us rewrite the Lagrangian given in Eq. (4) as follows:

$$L(X) = \sum_{k=1}^{K} \left(X_{.k}^{T} \mathcal{L} X_{.k} + \lambda (X_{\ell k} - Y_{.k})^{T} (X_{\ell k} - Y_{\ell k}) \right)$$

$$= \frac{1}{2} \sum_{k=1}^{K} \sum_{i,j=1}^{n} w_{ij} \left(\frac{X_{ik}}{\sqrt{d_i}} - \frac{X_{jk}}{\sqrt{d_j}} \right)^{2} + \lambda \sum_{k=1}^{K} \sum_{i=1}^{\ell} \left(X_{ik} - Y_{ik} \right)^{2}.$$

Thus, for all $k \in \{1, \ldots, K\}$, the first order condition $\dfrac{\partial L}{\partial X_{.k}}(X) = 0$ gives

$$\mathcal{L}X + \lambda S(X - Y) = 0,$$

where $S = \begin{pmatrix} I_{|\ell|} & 0 \\ 0 & 0 \end{pmatrix}$ is an $n \times n$ matrix. Using the block notation introduced earlier leads to the following equations:

$$\forall k \in \{1, \ldots, K\} : \mathcal{L}_{uu} X_{uk} + \mathcal{L}_{u\ell} X_{\ell k} = 0.$$

By recalling the condition $X_{\ell.} = Y_{\ell.}$, the last equation can be rewritten as

$$X_{u.} = -\mathcal{L}_{uu}^{-1} \mathcal{L}_{u\ell} Y_{\ell}. \tag{5}$$

Note that $\mathcal{L}_{uu}$ is an extracted block from the normalized Laplacian, hence is invertible, and the expression (5) is well defined as soon as each connected component of the graph has at least one labeled node. The expression (5) depends only on the value of the labeled nodes and on the topology of the graph.

3   Analysis on Random SBM Graphs

Let us set up the notations for SBM. Each node $i \in \{1, \ldots, n\}$ will belong to a cluster $\mathcal{C}_i$. Then, an edge is created between a pair of nodes (ij) with a probability that depends only on nodes' clusters:

$$\Pr\left((ij) \in E \right) = \mathrm{P}_{\mathcal{C}_i \mathcal{C}_j}.$$

The adjacency matrix A is thus a random matrix, whose expected value is

$$\mathbf{E}A_{ij} = \mathrm{P}_{\mathcal{C}_i \mathcal{C}_j}. \tag{6}$$

The weighted graph formed by the expected adjacency matrix of an SBM graph, given by (6), will be called mean field model.

It is common to call $p_i = \mathrm{P}_{\mathcal{C}_i \mathcal{C}_i}$ the intra-cluster edge probabilities and $q = \mathrm{P}_{\mathcal{C}_i \mathcal{C}_j}, i \neq j$ the inter-cluster edge probability (we assume that the inter-cluster edge probabilities are all equal to each other). We will denote by n_i the number of nodes in community i, with $n = \sum_{i=1}^{K} n_i$. Finally, d_i will be the average degree of nodes in cluster i.

We will mostly focus on the symmetric SBM with two communities, and will make use of the following assumptions; nonetheless, for each result, we will clearly state which assumption is needed. We think our results stand for more than two communities as well as in the non symmetric case (incorporating so-called Class Prior Knowledge, see for example [5] Section 10.8), to the price of harder and longer computations.

Assumption 1 (Symmetric SBM). *We consider an SBM graph with two communities of equal size $n_1 = n_2 = \frac{n}{2}$ and $p_1 = p_2 =: p$.*

Assumption 2 (Growing degrees). *The average degree d goes to infinity.*

Assumption 3 (Fixed fraction of labelled nodes). *The fraction of labeled nodes r remains constant as n grows to $+\infty$.*

Assumption 4 (Labeled nodes uniformly distributed). *Each community has the same fraction of labeled nodes (with respect to the community size), and they are chosen uniformly at random. Moreover, we assume that there is at least one labeled node in each connected component of the graph.*

The second part of Assumption 4 is needed to ensure that the extracted Laplacian $\mathcal{L}_{uu}$ is invertible. We can now state the main result of this paper.

Theorem 1 (Asymptotically almost exact recovery). *Label Spreading algorithm, defined by the minimization scheme (2)–(3), enables asymptotically almost exact recovery for an SBM graph under Assumptions 1–4.*

We will prove Theorem 1 in two steps. First, by doing exact calculation of the mean field solution X^{MF}, we will show that exact (even nonasymptotic) recovery is possible for the mean field model. Then, we will show that the solution of the minimization problem (2)–(3) is asymptotically concentrated sufficiently well around its mean field value. Those two results put together will give the proof of Theorem 1.

3.1   Exact Expression for Mean Field SBM

Recall that by mean field, we are referring to the situation where the random quantities are replaced by their means. In particular, we call mean field model the weighted graph formed by the expected adjacency matrix of an SBM graph.

In all the following, the subscript MF will be added to all quantities referring to the the mean field model. For simplicity of notations and computations, we will assume there is only two communities, but the analysis can be extended to K communities.

Let 1_{n_1} denote the column vector of size $n_1 \times 1$ with all entries equal to one, and by $J_{n_1;n_2} := 1_{n_1} 1_{n_2}^T$ the matrix of size $n_1 \times n_2$ with all entries equal to one. Furthermore, we will use a shorten notation J_{n_1} for $J_{n_1;n_1}$.

Without loss of generality and for the purpose of performance analysis, we implicitly assume that the first n_1 nodes are in cluster 1, whereas the last n_2 nodes are in cluster 2. Thus,

$$A^{MF} := \mathbf{E}A = \begin{pmatrix} p_1 J_{n_1} & q J_{n_1 n_2} \\ q J_{n_2 n_1} & p_2 J_{n_2} \end{pmatrix}.$$

In order for derivations to be more transparent, we also consider the case where diagonal elements of A^{MF} are not zero. This corresponds to a non-standard definition of SBM, where we could have edges $(i;i)$, with probability p_1 or p_2 depending on the community to whom i belongs to. Nonetheless, we could set the diagonals elements of A^{MF} to zero and our results would still hold.

Also without loss of generality and for the convenience of analysis, we will assume that the first rn_1 and the last rn_2 nodes are labeled. Note that if the quantities rn_i are not integers, we take their integer part, but we shall omit it to simplify the notations. Lastly, we introduce $\tilde{n}_i = (1 - r)n_i$ the number of unlabeled nodes in cluster i.

Theorem 2 (Exact expression for X^{MF}). *Let* $a = \dfrac{p_1}{d_1}$, $b = c = \dfrac{q}{\sqrt{d_1 d_2}}$,
$d = \dfrac{p_2}{d_2}$ *and* $F := \left(1 - \dfrac{p_1 \tilde{n}_1}{d_1}\right)\left(1 - \dfrac{p_2 \tilde{n}_2}{d_2}\right) - \tilde{n}_1 \tilde{n}_2 \dfrac{q^2}{d_1 d_2}$. *Then*

$$X_{u.}^{MF} = \begin{pmatrix} x_{11}^{MF} J_{(1-r)n_1} & x_{12}^{MF} J_{(1-r)n_1} \\ x_{21}^{MF} J_{(1-r)n_2} & x_{22}^{MF} J_{(1-r)n_2} \end{pmatrix},$$

where:

$$- \ x_{11}^{MF} = rn_1 \left(a - n_1 \frac{(1-r)a}{F}\left(-a + \tilde{n}_2(ad - bc)\right) + \frac{(1-r)bc}{F}n_2 \right);$$

$$- \ x_{12}^{MF} = rn_2 \left(b - \frac{(1-r)b}{F}n_1(-a + \tilde{n}_2(ad - bc)) + d\frac{(1-r)b}{F}n_2 \right);$$

$$- \ x_{21}^{MF} = rn_1 \left(c + rn_1 \frac{(1-r)ac}{F}n_1 - n_2 \frac{(1-r)c}{F}\left(-d + \tilde{n}_1(ad - bc)\right) \right);$$

$$- \ x_{22}^{MF} = rn_2 \left(d + \frac{(1-r)bc}{F}n_1 - n_2 \frac{(1-r)d}{F}(-d + \tilde{n}_1(ad - bc)) \right).$$

Proof. Recall from Eq. (5) that $X_{u.}^{MF} = -\left(\mathcal{L}_{uu}^{MF}\right)^{-1} \mathcal{L}_{u\ell}^{MF} Y_{\ell.}$.

First, let us notice that $\left(D^{-\frac{1}{2}} W D^{-\frac{1}{2}}\right)_{uu}^{MF} = \begin{pmatrix} a J_{\tilde{n}_1} & b J_{\tilde{n}_1 \tilde{n}_2} \\ c J_{\tilde{n}_2 \tilde{n}_1} & d J_{\tilde{n}_2} \end{pmatrix}$, where the quantities a, b, c and d are defined in the statement of the theorem. It follows from Proposition 2 in the Appendix that

$$\left(\mathcal{L}_{uu}^{MF}\right)^{-1} = I_{\tilde{n}} - \frac{1}{F}\left(\begin{pmatrix} -a + \tilde{n}_2(ad - bc) \end{pmatrix} J_{\tilde{n}_1} & -b J_{\tilde{n}_1 \tilde{n}_2} \\ -c J_{\tilde{n}_2 \tilde{n}_1} & \begin{pmatrix} -d + \tilde{n}_1(ad - bc) \end{pmatrix} J_{\tilde{n}_2} \right).$$

Moreover, $-\mathcal{L}_{u\ell}^{MF} = \begin{pmatrix} a J_{\tilde{n}_1;rn_1} & b J_{\tilde{n}_1;rn_2} \\ c J_{\tilde{n}_2;rn_1} & d J_{\tilde{n}_2;rn_2} \end{pmatrix}$ and $X_{\ell.} = \begin{pmatrix} 1_{rn_1} & 0_{rn_1} \\ 0_{rn_2} & 1_{rn_2} \end{pmatrix}$, thus

$$-\mathcal{L}_{u\ell}^{MF} X_{\ell.} = \begin{pmatrix} rn_1 a \, 1_{\tilde{n}_1} & rn_2 b \, 1_{\tilde{n}_1} \\ rn_1 c \, 1_{\tilde{n}_2} & rn_2 d \, 1_{\tilde{n}_2} \end{pmatrix},$$

and the product of $\left(\mathcal{L}_{uu}^{MF}\right)^{-1}$ by $-\mathcal{L}_{u\ell}^{MF} X_{\ell.}$ gives the stated result. $\qquad \square$

Proposition 1 (Exact recovery in mean field model). *The minimization procedure* (2)–(3) *achieves exact recovery in the mean field model of an SBM graph with two clusters of equal size, with $p_1 = p_2$ (Assumption 1), $p > q$ (associative communities) and with the same fraction $r > 0$ of labeled nodes in each cluster.*

Proof. Recall that the detection rule is given in Eq. (1). In the two communities case, recovery will be possible (and 100% correct) if and only if $x_{11} > x_{12}$ and $x_{22} > x_{21}$. By symmetry of the problem, it is enough to consider the condition $x_{11} > x_{12}$.

In the symmetric case, with two clusters of equal size ($n_1 = n_2$) and $p_1 = p_2$ (Assumption 1), it is then straightforward to see that

$$x_{11}^{MF} = r\frac{p}{p+q} + \frac{r(1-r)}{F}\frac{p}{(p+q)^2}(rp + (1-r)q) + \frac{r(1-r)}{F}\frac{q^2}{(p+q)^2},$$

$$x_{12}^{MF} = r\frac{q}{p+q} + \frac{r(1-r)}{F}\frac{q}{(p+q)^2}(rp + (1-r)q) + \frac{r(1-r)}{F}\frac{pq}{(p+q)^2}.$$

By subtracting those two lines, a little of algebra shows that

$$x_{11}^{MF} - x_{12}^{MF} = r\frac{p-q}{2q + r(p-q)}.$$

This last quantity is positive as soon as $p > q$, and this ends the proof.   □

We can make two remarks:

- First, note that we have not made any assumptions on the scaling of p_i and q with n, except that p_1 and p_2 are equal. In particular, the result holds in the case of logarithmic degree, which will be our main focus later on.
- Second, in the case of the mean field model, the result holds for finite n, thus it is exact (even non-asymptotic) recovery in the mean field model. It is not surprising, since recovery in the mean field model is obvious.

3.2   Concentration Towards Mean Field

Similarly to the concentration result in [3], we establish the concentration of X around its mean field value X^{MF} in terms of the Euclidean norm. For the sake of better readability, we omit the subscripts from $X_{.k}$ and $X_{.k}^{MF}$ ($k \in \{1, \ldots, K\}$) in the next theorem and the two following proofs. Similarly, we will shorten X_{uk} (respectively $Y_{\ell k}$) to X_u (respectively Y_ℓ).

Theorem 3. *Under the same assumptions as Theorem 1, for each class, the relative Euclidean distance between the solution X given by Label Spreading and its mean field value X^{MF} converges in probability to zero. More precisely, with high probability, we can find a constant $C > 0$ such that:*

$$\frac{\|X - X^{MF}\|}{\|X^{MF}\|} \leq \frac{C}{\sqrt{d}}. \tag{7}$$

Proof. Let us rewrite Eq. (5) as a perturbation of a system of linear equations corresponding to the mean field solution:

$$\left(\mathbf{E}\mathcal{L} + \Delta\mathcal{L}\right)_{uu}\left(X_u^{MF} + \Delta X_u\right) = -\left(\mathbf{E}\mathcal{L} + \Delta\mathcal{L}\right)_{u\ell} Y_\ell,$$

where $\Delta X := X - X^{MF}$ and $\Delta \mathcal{L} := \mathcal{L} - \mathbf{E}\mathcal{L}$.

Recall that a perturbation of a system of linear equations $(A + \Delta A)$ $(x + \Delta x) = b + \Delta b$ leads to the following sensitivity inequality (see e.g., Section 5.8 in [10]):

$$\frac{||\Delta x||}{||x||} \leq \frac{\kappa(A)}{1 - \kappa(A)\dfrac{||\Delta A||}{||A||}} \left(\frac{||\Delta b||}{||b||} + \frac{||\Delta A||}{||A||} \right)$$

where $||.||$ is a matrix norm associated to a vector norm $||.||$ (we used the same notations for simplicity) and $\kappa(A) := ||A^{-1}||.||A||$ the conditioning number. In our case, using spectral norm, this gives:

$$\frac{||X - X^{MF}||}{||X^{MF}||} \leq \frac{||\mathbf{E}\,\mathcal{L}_{uu}||.||(\mathbf{E}\,\mathcal{L}_{uu})^{-1}||}{1 - ||(\mathbf{E}\,\mathcal{L}_{uu})^{-1}||.||\Delta\,\mathcal{L}_{uu}||} \left(\frac{|| - \Delta\,\mathcal{L}_{u\ell}.Y_\ell||}{|| - \mathbf{E}\,\mathcal{L}_{u\ell}.Y_\ell||} + \frac{||\Delta\,\mathcal{L}_{uu}||}{||\mathbf{E}\,\mathcal{L}_{uu}||} \right).$$

Let us first deal with all the non random quantities. The spectral study of $\mathbf{E}\,\mathcal{L}_{uu}$ is done in the Appendix (Proposition 3). In particular, we have:

$$||\mathbf{E}\,\mathcal{L}_{uu}|| = \max\left\{ |\lambda| : \lambda \in \mathrm{Sp}(\mathbf{E}\,\mathcal{L}_{uu}) \right\} = 1,$$

$$\left\|(\mathbf{E}\,\mathcal{L}_{uu})^{-1}\right\| = \frac{1}{\min\left\{ |\lambda| : \lambda \in \mathrm{Sp}(\mathbf{E}\,\mathcal{L}_{uu}) \right\}} = \frac{1}{r}\frac{p+q}{p-q}.$$

Note that since p and q have the same dependency in n (from the assumptions, $p = a\dfrac{\log n}{n}$ and $q = b\dfrac{\log n}{n}$), the ratio $\dfrac{p+q}{p-q}$ does not depend on n, and $\left\|(\mathbf{E}\mathcal{L}_{uu})^{-1}\right\|$ is equal to a constant C'. We are left with the following inequality:

$$\frac{||X - X^{MF}||}{||X^{MF}||} \leq C'\frac{1}{1 - C'\,||\Delta\mathcal{L}_{uu}||} \left(\frac{||\Delta\mathcal{L}_{u\ell}.Y_\ell||}{||\mathbf{E}\,\mathcal{L}_{u\ell}.Y_\ell||} + ||\Delta\mathcal{L}_{uu}|| \right).$$

Moreover $\mathbf{E}\,\mathcal{L}_{u\ell}.Y_\ell = (1 - r)Y_u$, thus $||\mathbf{E}\,\mathcal{L}_{u\ell}.Y_\ell|| = (1-r)\sqrt{(1-r)n}$. So

$$\frac{||X - X^{MF}||}{||X^{MF}||} \leq \frac{C'}{1 - C'\,||\Delta\,\mathcal{L}_{uu}||} \left(\frac{||\Delta\,\mathcal{L}_{u\ell}||}{1 - r} + ||\Delta\,\mathcal{L}_{uu}|| \right), \tag{8}$$

where we used $||Y_\ell|| = \sqrt{rn}$ (since Y_ℓ is a vector of size rn with entries equal to 1 or -1) and $\sqrt{\dfrac{r}{1-r}} \leq 1$.

The concentration of the normalized Laplacian towards its mean field value has been established in [12]. In particular, the authors showed that w.h.p.

$$\left\|\mathcal{L} - \mathbf{E}\,\mathcal{L}\right\| = O\left(\frac{1}{\sqrt{d}}\right), \tag{9}$$

where d is the average degree, when $d = \Omega(\log n)$. However, the result of Eq. (9) is a concentration of the full normalized Laplacian (an $n \times n$ matrix), while here we are interested in concentration of an extracted matrix. Fortunately, concentration still holds, see Proposition 4 in the Appendix. Therefore, the terms $||\Delta \mathcal{L}_{uu}||$ and $||\Delta \mathcal{L}_{u\ell}||$ in Eq. (8) can be bounded by $\dfrac{K}{\sqrt{d}}$.

Last, C' being constant and $||\Delta \mathcal{L}_{uu}||$ going to zero, we can lower bound the term $\dfrac{C'}{1 - C' \, ||\Delta \mathcal{L}_{uu}||}$ by $2C'$ for n large enough, leaving us only with

$$\frac{||X - X^{MF}||}{||X^{MF}||} \leq \frac{C}{\sqrt{d}}$$

for a constant C. This ends the proof. $\qquad \square$

Inequality (7) indicates a slow convergence. For example, in the moderately sparse regime where $p(n)$ and $q(n)$ grows as a constant times $\dfrac{\log(n)}{n}$ (an interesting regime to study for SBM), we have established a bound on the convergence rate in the order of $\dfrac{1}{\sqrt{\log n}}$.

3.3   Asymptotically Almost Exact Recovery for SBM

Proof of Theorem 1. We just established a concentration inequality for X towards X^{MF}. In order to correctly classify a node i, one should hope that the node's value X_i is close enough to its mean field value X_i^{MF}. To be more precise, $|X_i - X_i^{MF}|$ should be smaller than half the community gap. Recall that in the symmetric case, we showed in Proposition 1 that the community gap is equal to $r\dfrac{p - q}{2q + r(p - q)}$, independent of n when p and q have the same dependency on n.

This leads us to define the notion of 'ϵ-bad nodes'. A node $i \in \{1, \ldots, n\}$ is said to be ϵ-bad if $|X_i - X_i^{MF}| > \epsilon$. Let us denote by B_ϵ the set of ϵ-bad nodes. The nodes that are not ϵ-bad, for an ϵ constant strictly smaller than half the community gap, are almost surely correctly classified.

From $||X - X^{MF}||^2 \geq \sum\limits_{i \in B_\epsilon} |X_i - X_i^{MF}|^2$, it comes that $||X - X^{MF}||^2 \geq |B_\epsilon| \times \epsilon^2$. Thus, using Theorem 3, we have w.h.p.:

$$|B_\epsilon| \leq \frac{C}{\epsilon^2} \frac{n}{d}. \tag{10}$$

If we take for ϵ a constant strictly smaller than half the community gap (recall that the community gap does not depend on n), then all nodes that are not in B_ϵ will be correctly classified. Since by (10) we have $|B_\epsilon| = o(n)$, the fraction of misclassified nodes is at most of order $o(1)$. This establishes almost exact recovery, and the proof of Theorem 1 is completed. $\qquad \square$

4   Discussion and Future Works

In this paper, we explicitly showed that Label Spreading can achieve good result, in the sense of almost exact recovery, for community detection on SBM graphs.

Our result stands in the case of two symmetric communities, but extension could be done for more than two non-symmetric communities, as well as labeled nodes non uniformly distributed across communities.

The case of sub-linear number of labeled nodes is worthy of further investigation. As was noted in [13], semi-supervised methods like Label Spreading tend to fail in the limit of small labeled data. Indeed, the minimization scheme (2)–(3) rely too heavily on the condition $X_\ell = Y_\ell$ and not enough on the graph structure. For example, in the extreme case where r is equal to zero, then the solution X have all entries equal, and recovery is not possible. But in that case, we should aim to recover the solution given by unsupervised Spectral Clustering method. Such modified versions of Label Spreading could be part of future research, and should greatly improve the results (at least in the limit of r going to zero).

In particular, we could see if such improved methods could achieve exact recovery under weaker conditions than unsupervised methods. It was shown that unsupervised methods can recover the exact community structure of SBM when $p = a\dfrac{\log n}{n}$ and $q = b\dfrac{\log n}{n}$ if and only if $\dfrac{a+b}{2} > 1 + \sqrt{ab}$. Since $\dfrac{a+b}{2} > 1$ is the connectivity requirement for a symmetric SBM, we can see that connectivity is required (as expected), but not sufficient. Lowering this bound in the semi-supervised scenario, and be able to remove this $\sqrt{ab}$ oversampling factor, would be an interesting result, as we would have exact recovery with semi-supervised setting if and only if the SBM graph is connected.

Acknowledgements. This work has been done within the project of Inria – Nokia Bell Labs "Distributed Learning and Control for Network Analysis".

A   Background Results on Matrix Analysis

A.1   Inversion of the Identity Matrix Minus a Rank 2 Matrix

Lemma 1 (Sherman-Morrison-Woodbury formula). *Let A be an invertible $n \times n$ matrix, and B, C, D matrices of correct sizes. Then:* $\left(A + BCD\right)^{-1} = A^{-1} - A^{-1}B\left(I + CDA^{-1}B\right)^{-1}CDA^{-1}$. *In particular, if u, v are two column vectors of size $n \times 1$, we have:* $\left(A + uv^T\right)^{-1} = A^{-1} - \dfrac{A^{-1}uv^T A^{-1}}{1 + v^T A^{-1}u}$.

Proof. See for example [10], section 0.7.4.   □

Lemma 2. *Let $M = \begin{pmatrix} aJ_{n_1} & bJ_{n_1n_2} \\ cJ_{n_2n_1} & dJ_{n_2} \end{pmatrix}$ for some values a, b, c, d. Let $n = n_1 + n_2$. If $I_n - M$ is invertible, we have:*

$$(I - M)^{-1} = I_n - \frac{1}{K}\begin{pmatrix} \left(-a + n_2(ad - bc)\right)J_{n_1} & -bJ_{n_1n_2} \\ -cJ_{n_2n_1} & \left(-d + n_1(ad - bc)\right)J_{n_2} \end{pmatrix}$$

where $K = (1 - n_1 a)(1 - n_2 d) - n_1 n_2 bc$.

Proof. We will use the Sherman-Morrison-Woodbury matrix identity (Lemma 1) with $A = I_n$, $D = \begin{pmatrix} 1 \ldots 1; 0 \ldots 0 \\ 0 \ldots 0; 1 \ldots 1 \end{pmatrix}$ (on the first line, there are n_1 ones and n_2 zeros), $B = D^T$ and $C = \begin{pmatrix} -a & -b \\ -c & -d \end{pmatrix}$. We can easily verify that $BCD = -M$.

$$(I - M)^{-1} = I_n - B(I + CDB)^{-1}CD$$
$$= I_n - B\begin{pmatrix} 1 - n_1 a & -n_2 b \\ -n_1 c & 1 - n_2 d \end{pmatrix}^{-1} CD$$
$$= I_n - B \frac{1}{(1 - n_1 a)(1 - n_2 d) - n_1 n_2 bc} \begin{pmatrix} 1 - n_2 d & n_2 b \\ n_1 c & 1 - n_1 a \end{pmatrix} CD$$
$$= I_n - \frac{1}{K} B \begin{pmatrix} -a + n_2(ad - bc) & -b \\ -c & -d + n_1(ad - bc) \end{pmatrix} D$$
$$= I_n - \frac{1}{K} \begin{pmatrix} \left(-a + n_2(ad - bc)\right) J_{n_1} & -b J_{n_1 n_2} \\ -c J_{n_2 n_1} & \left(-d + n_1(ad - bc)\right) J_{n_2} \end{pmatrix}.$$

□

A.2   Spectral Study of a Rank 2 Matrix

Lemma 3 (Schur's determinant identity, [10]). *Let A, D and $\begin{pmatrix} A & B \\ C & D \end{pmatrix}$ be squared matrices. If A is invertible, we have:*

$$\det \begin{pmatrix} A & B \\ C & D \end{pmatrix} = \det(A) \det(D - CA^{-1}B).$$

Proof. Follows from the formula $\begin{pmatrix} A & B \\ C & D \end{pmatrix} = \begin{pmatrix} A & 0 \\ C & I_q \end{pmatrix} \begin{pmatrix} I_p & A^{-1}B \\ 0 & D - CA^{-1}B \end{pmatrix}.$ □

Lemma 4 (Matrix determinant lemma, [10]). *For an invertible matrix A and two column vectors u and v, we have $\det(A + uv^T) = (1 + v^T A^{-1} u) \det(A)$.*

Lemma 5. *Let α and β be two constants. When $M = \alpha I_n + \beta J$ where J is the $n \times n$ matrix with all entries equal to one, we have $\det M = \alpha^{n-1}(\alpha + \beta n)$.*

Proof. Suppose that $\alpha \neq 0$. Then with $v^T = (1, \ldots, 1)$ and $u = \beta(1, \ldots, 1)$ vectors of size $1 \times n$, Lemma 4 gives us

$$\det M = \det(\alpha I_n)\left(1 + v^T(\alpha I_n)^{-1} u\right)$$
$$= \alpha^n \left(1 + \frac{\beta n}{\alpha}\right)$$
$$= \alpha^{n-1}(\alpha + \beta n),$$

which proves the lemma for $\alpha \neq 0$. To treat the case $\alpha = 0$, see that the function $\alpha \in \mathbf{R} \mapsto \det(\alpha I_n + \beta J)$ is continuous (even analytic) [4], thus by continuous prolongation in $\alpha = 0$, the expression $\alpha^{n-1}(\alpha + \beta n)$ holds for any $\alpha \in \mathbf{R}$.   □

Proposition 2. *Let* $M = \begin{pmatrix} aJ_{n_1} & bJ_{n_1 n_2} \\ cJ_{n_2 n_1} & dJ_{n_2} \end{pmatrix}$ *for some values* a, b, c, d. *The eigenvalues of* M *are:*

- *0 with multiplicity* $n_1 + n_2 - 2$;
- $\lambda_\pm = \dfrac{1}{2}\left(n_1 a + n_2 d \pm \sqrt{\Delta}\right)$ *where* $\Delta = (n_1 a - n_2 d)^2 + 4 n_1 n_2 bc$.

Proof. The matrix being of rank 2 (except for some degenerate cases), the fact that 0 is an eigenvalue of multiplicity $n_1 + n_2 - 2$ is obvious. By an explicit computation of the characteristic polynomial of M, the two remaining eigenvalues will be given as roots of a polynomial of degree 2.

Let $\lambda \in \mathbf{R}$ and $A := \lambda I_{n_1} - aJ_{n_1}$. If $\lambda \notin \{0; an_1\}$, then A is invertible and by the Schur's determinant identity (Lemma 3) we have

$$\det(\lambda I_n - M) = \det A \, \det\left(\lambda I_{n_2} - dJ_{n_2} - cJ_{n_2 n_1} A^{-1} bJ_{n_1 n_2}\right)$$
$$= \det A \, \det B.$$

From Lemma 5, it follows that $\det A = \lambda^{n_1 - 1}(\lambda - n_1 a)$.

Let us now compute $\det B$. First, we show that $A^{-1} = \dfrac{1}{\lambda}\left(I_{n_1} + \dfrac{a}{\lambda - an_1} J_{n_1}\right)$. Indeed, from the Sherman-Morrison-Woodbury formula (Lemma 1) with $u = -a1_{n_1}$ and $v = 1_{n_1}$, it follows that

$$\left(\lambda I_{n_1} - aJ_{n_1}\right)^{-1} = \frac{1}{\lambda} I_{n_1} - \frac{1}{\lambda^2} \frac{-aJ_{n_1}}{1 + \dfrac{-an_1}{\lambda}}$$
$$= \frac{1}{\lambda} I_{n_1} + \frac{1}{\lambda} \frac{a}{\lambda - an_1} J_{n_1},$$

which gives the desired expression. Thus,

$$B = \lambda I_{n_2} - dJ_{n_2} - \frac{bc}{\lambda} J_{n_2 n_1}\left(I_{n_1} + \frac{a}{\lambda - an_1} J_{n_1}\right) J_{n_1 n_2}$$
$$= \lambda I_{n_2} - dJ_{n_2} - \frac{bc}{\lambda}\left(n_1 + \frac{a\, n_1^2}{\lambda - an_1}\right) J_{n_2}$$
$$= \lambda I_{n_2} + \left(-d - \frac{bcn_1}{\lambda - an_1}\right) J_{n_2}.$$

Again, this matrix is of the form $\lambda I_n + \beta J_n$, and we can use Lemma 5 to show that

$$\det B = \lambda^{n_2 - 1}\left(\lambda + n_2 \beta\right).$$

Now we can finish the computation of $\det(\lambda I_n - M)$

$$\det(\lambda I_n - M) = \lambda^{n_1 + n_2 - 2}(\lambda - n_1 a)\left(\lambda - n_2 d - \frac{bcn_1 n_2}{\lambda - an_1}\right)$$
$$= \lambda^{n_1 + n_2 - 2}\left(\lambda^2 + \lambda(-n_1 a - n_2 d) + n_1 n_2 (ad - bc)\right).$$

The discriminant of this second degree polynomial expression is given by

$$\Delta = (n_1 a + n_2 d)^2 - 4n_1 n_2 (ad - bc)$$
$$= (n_1 a - n_2 d)^2 + 4n_1 n_2 bc.$$

Thus $\Delta \geq 0$ and the two remaining eigenvalues are given by

$$\lambda_\pm = \frac{1}{2}\left(n_1 a + n_2 d \pm \sqrt{\Delta}\right).$$

$\square$

A.3   Spectral Study of $\mathbf{E}\mathcal{L}$

Proposition 3 (Eigenvalues of $\mathbf{E}\mathcal{L}_{uu}$, symmetric case). *Assume two communities of equal size, with $p_1 = p_2 (= p)$. The two smallest eigenvalues of $\mathbf{E}\mathcal{L}_{uu}$ are:*

$$\lambda_1 = r\frac{p-q}{p+q} \quad and \quad \lambda_2 = r.$$

Note that the other eigenvalue of $\mathbf{E}\mathcal{L}_{uu}$ is one (with multiplicity $\lfloor (1-r)n \rfloor - 2$).

Proof. The matrix $\mathbf{E}\mathcal{L}_{uu}$ can be written as $I - M$, where $M = D^{-1/2}AD^{-1/2}$ has a block form like in Proposition 2, with coefficients $a = \frac{p_1}{d_1}$, $b = c = \frac{q}{\sqrt{d_1 d_2}}$ and $d = \frac{p_2}{d_2}$. Note that the blocks sizes are now $\lfloor (1-r)n_i \rfloor$ and not n_i. Under the symmetric assumption, we have $d_1 = d_2 = \frac{n}{2}(p+q)$.

Moreover, λ_M is an eigenvalue of M if and only if $1 - \lambda_M$ is eigenvalue of $\mathbf{E}\mathcal{L}_{uu}$. Using the notations of Proposition 2, we have $\Delta = 4(1-r)^2 \frac{q^2}{(p+q)^2}$, and the two non-zero eigenvalues of M are given by:

$$\lambda_\pm = \frac{1}{2}\left(2(1-r)\frac{p}{p+q} \pm 2(1-r)\frac{q}{p+q}\right)$$
$$= 1 - r\frac{p \pm q}{p+q}.$$

$\square$

B   Spectral Norm of an Extracted Matrix

Proposition 4. *Let A be a matrix and B an extracted matrix (non necessarily squared: we can remove rows and columns with different indices, and potentially more rows than columns, or vice versa) from A, then: $\|B\|_2 \leq \|A\|_2$.*

Proof. For two subsets I and J of $\{1, \ldots, n\}$, let $B = A_{IJ}$ the matrix obtained from A by keeping only the rows (resp. columns) in I (resp. in J). Then $B = M_1 A M_2$ where M_1 and M_2 are two appropriately chosen permutation matrices. Thus their spectral norm is equal to one, and the result $\|B\|_2 \leq \|A\|_2$ follows from the inequality $\|B\|_2 \leq \|M_1\|_2 \|A\|_2 \|M_2\|_2$. $\square$

References

1. Abbe, E.: Community detection and stochastic block models. Found. Trends® Commun. Inf. Theory **14**(1–2), 1–162 (2018)
2. Avrachenkov, K., Gonçalves, P., Mishenin, A., Sokol, M.: Generalized optimization framework for graph-based semi-supervised learning. In: SIAM International Conference on Data Mining (SDM 2012) (2012)
3. Avrachenkov, K., Kadavankandy, A., Litvak, N.: Mean field analysis of personalized pagerank with implications for local graph clustering. J. Stat. Phys. **173**(3–4), 895–916 (2018)
4. Avrachenkov, K.E., Filar, J.A., Howlett, P.G.: Analytic Perturbation Theory and Its Applications, vol. 135. SIAM, Philadelphia (2013)
5. Chapelle, O., Schölkopf, B., Zien, A.: Semi-supervised Learning. Adaptive Computation and Machine Learning. MIT Press, Cambridge (2006)
6. Condon, A., Karp, R.M.: Algorithms for graph partitioning on the planted partition model. In: Hochbaum, D.S., Jansen, K., Rolim, J.D.P., Sinclair, A. (eds.) APPROX/RANDOM-1999. LNCS, vol. 1671, pp. 221–232. Springer, Heidelberg (1999). https://doi.org/10.1007/978-3-540-48413-4_23
7. Erdős, P., Rényi, A.: On random graphs. Publ. Math. (Debr.) **6**, 290–297 (1959)
8. Gilbert, E.N.: Random graphs. Ann. Math. Statist. **30**(4), 1141–1144 (1959)
9. Holland, P.W., Laskey, K.B., Leinhardt, S.: Stochastic blockmodels: first steps. Soc. Netw. **5**(2), 109–137 (1983)
10. Horn, R.A., Johnson, C.R.: Matrix Analysis, 2nd edn. Cambridge University Press, Cambridge (2012)
11. Johnson, R., Zhang, T.: On the effectiveness of Laplacian normalization for graph semi-supervised learning. J. Mach. Learn. Res. **8**(Jul), 1489–1517 (2007)
12. Le, C.M., Levina, E., Vershynin, R.: Concentration and regularization of random graphs. Random Struct. Algorithms **51**(3), 538–561 (2017)
13. Mai, X., Couillet, R.: A random matrix analysis and improvement of semi-supervised learning for large dimensional data. J. Mach. Learn. Res. **19**(1), 3074–3100 (2018)
14. Zhou, D., Bousquet, O., Lal, T.N., Weston, J., Schölkopf, B.: Learning with local and global consistency. In: Advances in Neural Information Processing Systems, pp. 321–328 (2004)
15. Zhu, X.: Semi-supervised learning literature survey. Technical report, Computer Science Department, University of Wisconsin-Madison (2006)
16. Zhu, X., Ghahramani, Z., Lafferty, J.D.: Semi-supervised learning using Gaussian fields and harmonic functions. In: ICML (2003)

Strongly n-e.c. Graphs and Independent Distinguishing Labellings

Christopher Duffy[1]([✉]) and Jeannette Janssen[2]

[1] Department of Mathematics and Statistics, University of Saskatchewan,
Saskatoon, Canada
`christopher.duffy@usask.ca`
[2] Department of Mathematics and Statistics, Dalhousie University, Halifax, Canada

Abstract. A countable graph G is n-ordered if its vertices can be enumerated so each vertex has no more than n neighbours appearing earlier in the enumeration. Here we consider both deterministic and probabilistic methods to produce n-ordered countable graphs with universal adjacency properties. In the countably infinite case, we show that such universal adjacency properties imply the existence an independent 2-distinguishing labelling.

Keywords: Graph evolution · n-ordered graphs · Graph distinguishing

1 Introduction

Complex networks are set apart within the class of graphs by features that are representative of real-world networks but do not generally appear within Erdős-Rényi random graphs [16]. For example many real-world networks are *small-world* [24]. That is, they exhibit relatively small average shortest path lengths, but higher clustering coefficients than are seen in Erdős-Rényi graphs with similar average shortest path length. Real-world networks also often differ from Erdős-Rényi graphs in the distribution of vertex degree [6]. Erdős-Rényi graphs have a relatively uniform degree distribution, whereas many real-world networks have power-law distributed degree distribution. That is, complex networks are often *scale-free*. As a consequence of such differences, a number of graph construction models (both deterministic and stochastic) have been developed so that the resulting graphs are better representative of real-world networks [15,18,25,26].

One well-studied class of complex network models are *Preferential Attachment* models [1,2,21,22]. At each step in an online process, new vertices are added so that they are more likely to be adjacent to existing vertices of high degree than existing vertices of low degree. One particular application of Preferential Attachment models is modelling the growth of the World Wide Web [5].

Complex networks generated with a online process often proceed through many phases where stopping the process at any point within a particular phase

© Springer Nature Switzerland AG 2019
K. Avrachenkov et al. (Eds.): WAW 2019, LNCS 11631, pp. 44–56, 2019.
https://doi.org/10.1007/978-3-030-25070-6_4

will produce a graph whose parameters are expected to be contained within a particular range. And so such generation processes may be used to generate complex networks whose features can be finely tuned by both the initial parameters and the length of the process. For example the Watts-Strogatz small-world graph model [26] can be used to generate random graphs whose expected clustering coefficient can be varied as a function of an input parameter, but whose expected average path length scales linearly with the length of the process. In recent years the limiting behaviour of such models have received attention from both pure and applied researchers [7,9].

Consider the graph formed by the following online process. The graph G_0 consists of a single vertex. For all $i > 0$, the finite graph G_i is formed from G_{i-1} by adding a dominating vertex to each of the subsets of $V(G_{i-1})$. Surprisingly, the graph formed as $i \to \infty$ is isomorphic the *infinite random*, or *Rado*, graph [11]. That is, it is isomorphic to the graph with a countably infinite set of vertices, where each pair of vertices is adjacent with some fixed probability $p \in (0,1)$.

For a graph G and a pair of disjoint of subsets $A, B \subset V(G)$ a vertex of $V(G) \setminus (A \cup B)$ is *correctly joined to A and B* if it adjacent to every vertex in A and no vertex in B. A graph is *n-existentially closed* (*n-e.c.*) if for every pair of disjoint subsets $A, B \subset V(G)$ so that $|A \cup B| \le n$ there is a vertex correctly joined to A and B. Though for any fixed n, nearly every graph is n-e.c. [16], constructing explicit examples of n-e.c. graphs and graph families it a difficult task utilizing a wide variety of combinatorial tools. We direct the reader to a survey of construction techniques for n-e.c. graphs and graph families [8]. The infinite random graph R is characterized as being the unique infinite graph that is n-e.c. for all n. We direct the reader to a comprehensive resource on various aspects of R [11].

In this paper we study the limiting behaviour of the process above when we vary the input graph and put a fixed upper bound on the size of the subsets that are dominated. As this process proceeds through discrete time-steps the generated graphs seemingly display features of both scale-free and small-world networks. We fully classify the limiting behaviour of these processes and show that these deterministic processes can be modelled by a stochastic process. We conclude with an application of our work in the study of graph distinguishing.

In [9] the authors consider a variation of the deterministic construction of R by letting $G_0 = K_n$ (for some fixed n) and forming G_{i-1} from G_i by adding a dominating vertex to every subset of vertices of order exactly n in $V(G_{i-1})$. The resulting infinite graph formed as $t \to \infty$ is denoted $R^{(n)}$. The infinite graph $R^{(n)}$ satisfies a stronger existential closure property: A graph G is *strongly n-existentially closed* if every pair of finite disjoint subsets $A, B \subset V(G)$ so that $|A| \le n$ there is a vertex correctly joined to A and B. That there is no restriction on the order of B implies directly that for every n, every strongly n-existentially closed graph is infinite.

Along with the deterministic construction of $R^{(n)}$ akin to that of R in [9], the authors present a probabilistic construction. They further introduce the graph

$R^{(H,n)}$, the graph formed in the same manner as $R^{(n)}$ modified so that $G_0 = H$, a fixed graph on n vertices, but leave open the question of when $R^{(G,n)} \cong R^{(H,n)}$. In this article we answer this question, and show that the tools and techniques used in [9] may be used to study a wider class of strongly n-e.c. graphs.

Let G and H be graphs. A homomorphism of G to H is a vertex mapping that preserves adjacency. That is, $\phi : V(G) \to V(H)$ is a homomorphism when for all $uv \in E(G)$ we have $\phi(u)\phi(v) \in E(H)$. If ϕ is such a mapping, then we write $\phi : G \to H$. For a homomorphism $\phi : G \to H$ and $X \subseteq V(G)$, let $\phi(X) = \{y : \phi(x) = y, x \in X\}$. If $X = V(G)$, denote $\phi(X)$ by $Im(\phi)$. A bijective homomorphism is an *isomorphism*. A homomorphism $\phi : G \to G$ is an *automorphism* when $Im(\phi) = V(G)$.

For $S \subseteq V(G)$, we denote by $G[S]$ the subgraph of G induced by the vertices of S. When there is no possibility for confusion, we write $[S]$ rather than $G[S]$. If H is a subgraph of G, we write $H \leq G$ and we let $G - H = G[V(G) \setminus V(H)]$. If H is a countably infinite graph with vertex set $u_1, u_2, \ldots$, then let $H_i^* = H[\{u_1, u_2, \ldots, u_i\}]$. That is, H_i^* is the subgraph induced by the first i vertices of the enumeration.

Following [9], we define n-*ordered*. For fixed n, G is n-*ordered* if there exists a well-ordering of its vertices $(x_i : i \in I)$, where $|I|$ is finite or I has the order-type $\mathbb{N}$ so that each x_j has at most n neighbours x_i with $i < j$. The ordering $(x_i, i \in I)$ is an n-*ordering* of $V(G)$.

Let $n_G = |V(G)|$ and $e_G = |E(G)|$. Herein we assume that all graphs are countable and simple. For all other definitions and notations we refer the reader to [10].

Denote by $R^{(H,n)}$ the graph formed from the limit as $t \to \infty$ of the following process. Let $H_0 = H$ be a finite graph on at least n vertices. The graph H_i is constructed from H_{i-1} as follows: For each $S \subseteq V(H_{i-1})$ such that $|S| = n$ we add a vertex x_S with neighbourhood S. We say that x_S (n)-*extends* S and that x_S is an (n)-extension of S. Let X_i be the set of vertices contained in $H_i - H_{i-1}$. Observe that X_i is an independent set in H_i and each vertex of X_i has degree exactly n in H_i. The vertices of $R^{(H,n)}$ may be enumerated using the sequence $V(H), X_1, X_2, \ldots$. Given some fixed ordering of the vertices of H, order the vertices of X_1 in dictionary order based on the indices of their neighbours in H_0. Repeating this process inductively for each X_i yields a vertex ordering of $R^{(H,n)}$ so that the first n_H vertices of the ordering induce a copy and H, and each of the subsequent vertices has exactly n vertices earlier in the ordering. This vertex ordering is fully determined by the choice of the ordering chosen for H. We call such an ordering a *natural* (n)-*enumeration*. When there is no possibility for confusion we call such an ordering a natural enumeration. Observe that if H is n-ordered and the vertices of H are fixed in an n-ordering, then natural (n)-enumeration of the vertices of $R^{(H,n)}$ is an n-ordering of the vertices of $R^{(H,n)}$.

Denote by $R^{(H,\leq n)}$ the countably infinite graph formed from the limit as $t \to \infty$ of the following process. Let $H_0 = H$ be a graph. The graph H_i is constructed from H_{i-1} as follows: For each $S \subseteq V(H_{i-1})$ such that $|S| \leq n$ add a vertex x_S with neighbourhood S. We say that x_S $(\leq n)$-*extends* S and that x_S

is an $(\leq n)$-extension of S. Let X_i be the set of vertices contained in $H_i - H_{i-1}$. Observe that X_i is an independent set in H_i and each vertex of X_i has degree at most n in H_i. The vertices of $R^{(H, \leq n)}$ may be enumerated using the sequence $V(H), X_1, X_2, \ldots$. Given some fixed ordering of the vertices of H, order the vertices of X_1 in dictionary order based on the indices of their neighbours in H_0. Repeating this process inductively for each X_i yields a vertex ordering of $R^{(H, \leq n)}$ so that the first n_H vertices of the ordering induce a copy and H, and each of the subsequent vertices has at most n vertices earlier in the ordering. This vertex ordering is fully determined by the choice of the ordering chosen for H. We call such an ordering a *natural* $(\leq n)$-*enumeration* of the vertices of $R^{(H, \leq n)}$. When there is no possibility for confusion we call such an ordering a natural enumeration. Observe that if H is n-ordered and the vertices of H are fixed in an n-ordering, then a natural $(\leq n)$-enumeration of the vertices of $R^{(H, \leq n)}$ is an n-ordering of the vertices of $R^{(H, \leq n)}$.

For a fixed finite graph H, let $u_1, u_2, \ldots$ be a natural enumeration of the vertices of $R^{(H, n)}$ or of $R^{(H, \leq n)}$ given by the sequence $V(H), X_1, X_2, \ldots$ as described above. For a pair of vertices u_i, u_j with $j > n_{H_0}$ we say that u_i is an *ancestor* of x_j and u_j is a descendant of u_i when there is a finite path $u_i u_{k_1} u_{k_2} \ldots u_{k_\ell} u_j$ such that $i < k_1 < k_2 < \cdots < k_\ell < j$. We denote by $\text{age}(u_i)$ the index t such that $u_i \in X_t$ when $i > n_H$. Otherwise $\text{age}(u_i) = 0$. For a finite subset of vertices A, let $\text{age}(A) = \max_{v \in A}\{\text{age}(v)\}$. Observe that if u_i is an ancestor of u_j, then necessarily $\text{age}(u_i) < \text{age}(u_j)$.

In this work we show how the tools and methods used in [9] to study $R^{(n)}$ may be extended and modified to study $R^{(H, n)}$ and $R^{(H, \leq n)}$. In particular, we show that though $R^{(G, n)} \not\cong R^{(H, \leq n)}$ for any pair of finite graphs G and H, countably infinite graphs of the form $R^{(G, n)}$ and $R^{(H, \leq n)}$ exhibit many of the same properties as $R^{(n)}$, including a probabilistic construction. We fully answer the question posed in [9] of necessary and sufficient conditions such that $R^{(n)} \cong R^{(H, n)}$. In the main result of this work we further extend these conditions to give necessary and sufficient conditions for when $R^{(G, n)} \cong R^{(H, n)}$ and when $R^{(G, \leq n)} \cong R^{(H, \leq n)}$.

To study graphs formed by these two processes, we require the following relations. For a pair of countable graphs H_0 and H we write $H_0 \prec_n H$ if H can be formed from H_0 by iteratively adding vertices of degree n. We write $H_0 \prec_{\leq n} H$, if H can be formed from H_0 by iteratively adding vertices of degree at most n. Note that by definition we have $H \prec_n R^{(H, n)}$ and $H \prec_{\leq n} R^{(H, \leq n)}$.

Theorem 1. *The relations $\prec_n$ and $\prec_{\leq n}$ are transitive on countable graphs.*

The proof of Theorem 1 is given in [9] for $\prec_n$. The proof that $\prec_{\leq n}$ is transitive proceeds similarly to that of $\prec_n$, and thus is omitted.

2   Constructing Infinite Graphs by (n)-extensions

In [9] the authors give the following two results for $R^{(n)}$:

Theorem 2 ([9]). *Let Γ be a countably infinite graph. We have $\Gamma \cong R^{(n)}$ if and only if Γ is strongly n-e.c. and $K_n \prec_n \Gamma$.*

Theorem 3 ([9]). *$R^{(n)}$ is strongly n-e.c., but not strongly $n+1$-e.c.*

The proofs of each of these results easily extend to $R^{(H,n)}$.

Theorem 4. *Let H be a finite graph with atleast n vertices. Let Γ be a countably infinite graph. We have $\Gamma \cong R^{(H,n)}$ if and only if Γ is strongly n-e.c. and $H \prec_n \Gamma$.*

Theorem 5. *$R^{(H,n)}$ is strongly n-e.c., but not strongly $n+1$-e.c.*

Together these conditions give a sufficient condition for a countably infinite graph to be strongly n-e.c., but not strongly $n+1$-e.c..

Corollary 1. *Let Γ be a strongly n-e.c. graph. If there exists a finite graph H such that $H \prec_n \Gamma$, then Γ is not strongly $n+1$-e.c..*

Using these theorems, we provide necessary results to study the question of when $R^{(G,n)} \cong R^{(H,n)}$.

Lemma 1. *Let G be a graph on at least n vertices. Let $S \subseteq V(G)$ such that $|S| = n$. If G' is the graph formed from G by adding a single vertex x adjacent to each vertex in S, then $R^{(G,n)} \cong R^{(G',n)}$.*

The proof of Lemma 1 proceeds via a *back and forth* argument to construct a sequence of partial isomorphisms $\phi_i : R^{(G,n)} \to R^{(G',n)}$. The limit of this sequence is an isomorphism $\phi : R^{(G,n)} \to R^{(G',n)}$.

Theorem 6. *If H and G are finite graphs so that $H \prec_n G$, then $R^{(H,n)} \cong R^{(G,n)}$.*

Proof. Let $H_0 = H$ and $G_0 = G$. If $H_0 \prec_n G_0$, then there exist a sequence of finite graphs: $H_0 < H_1 < H_2 < \cdots < H_t \cong G_0$ such that for each pair H_i, H_{i-1} there exists $S \subseteq V(H_{i-1})$ and a vertex $x \in V(H_i)$ so that H_i is formed from H_{i-1} by adding x and edges between x and every vertex of S. By Lemma 1 we have $R^{(H_i,n)} \cong R^{(H_{i-1},n)}$ for all $1 \le i \le t$. The result follows from the transitivity of isomorphism and the transitivity of $\prec_n$. $\qquad \square$

Using Theorem 6, we give necessary and sufficient conditions such that $R^{(G,n)} \cong R^{(H,n)}$.

Theorem 7. *Let G and H be finite graphs. We have $R^{(G,n)} \cong R^{(H,n)}$ if and only if there exists a finite graph K such that $G \prec_n K$ and $H \prec_n K$.*

Proof. Assume there exists a finite graph K such that $G \prec_n K$ and $H \prec_n K$. By Theorem 6 and the transitivity of isomorphism it follows directly that $R^{(G,n)} \cong R^{(H,n)}$.

Assume $R^{(G,n)} \cong R^{(H,n)}$. Note that we may assume $n_G = n_H$, as if $n_G < n_H$, then we can construct G' so that $G \prec_n G'$ and $n_{G'} = n_H$. Let $H_0 = H$ and $G_0 = G$. Let $\Gamma_H = R^{(H_0,n)}$ and $\Gamma_G = R^{(G_0,n)}$. Let $y_1, y_2, \ldots$ be a natural enumeration of the vertices of Γ_H. Similarly, let $z_1, z_2, \ldots$ be a natural enumeration of the vertices of Γ_G. Let $\phi : \Gamma_H \to \Gamma_G$ be an isomorphism. Since ϕ is an isomorphism there exists $A \subset V(\Gamma_G)$ such that $\phi^{-1}(A) = \{y_1, y_2, \ldots, y_{n_H}\}$. Let $A_0 = A$. For $k > 0$ let A_k be the union of A_{k-1} and those vertices z_j that have a neighbour $z_i \in A_{k-1}$ such that $i > j$. Observe that since only vertices with an index smaller than the maximum of those in A_{k-1} are added to form A_k, there exists some smallest t such that $A_t = A_{t+1}$.

Let K_G be the subgraph induced by the vertices of $A_t \cup \{z_1, \ldots, z_{n_G}\}$. Notice that by construction we have $G_0 \prec_n K_G$. Let $X = A_t \setminus (A_0 \cup \{z_1, \ldots, z_{n_{G_0}}\})$. That is, X is the set of vertices of A contained neither in the initial copy of G_0 in Γ_G nor in the image of H_0 under ϕ. Observe $e_{K_G} = e_{G_0} + n(|X| + n_{H_0})$. Since $n_{G_0} = n_{H_0}$, we have $e_{K_G} = e_{G_0} + n(|X| + n_{G_0})$.

Consider $\phi^{-1}(V(K_G))$. Let $K_H = [\phi^{-1}(V(K_G))]$. If $H_0 \prec_n K_H$, then the proof is complete as ϕ is an isomorphism and $K_H \cong K_G$. And so assume $H_0 \not\prec_n K_H$. Observe K_H is a subgraph of Γ_H and H_0 is an induced subgraph of K_H. Therefore there is an ordering of the vertices of K_H, $u_1, u_2, \ldots$, so that $[u_1, u_2, \ldots, u_{n_{H_0}}] \cong H_0$ and for each $i > n_{H_0}$ the vertex u_i has no more than n neighbours appearing earlier in the ordering. Since $H_0 \not\prec_n K_H$ there is a vertex u_j with $j > n_{H_0}$ so that u_j has strictly fewer than n neighbours appearing earlier in the ordering. Therefore $e_{K_H} < e_{H_0} + n(|X| + n_{G_0})$.

Since $e_{K_H} = e_{K_G}$ we have $e_{H_0} + n(|X| + n_{G_0}) < e_{G_0} + n(|X| + n_{G_0})$ and so $e_{H_0} < e_{G_0}$. However, one may construct a similar argument using an isomorphism $\phi' : \Gamma_G \to \Gamma_H$ to show $e_{G_0} < e_{H_0}$. This is a contradiction. And so if $R^{(G_0,n)} \cong R^{(H_0,n)}$, then there exists a finite graph K so that $G_0 \prec_n K$ and $H_0 \prec_n K$. □

Corollary 2. *Let G and H be finite graphs on at least n vertices. If $R^{(G,n)} \cong R^{(H,n)}$, then there exists a constant c so that $e_G - e_H = cn$*

Corollary 3. *For every $n \geq 1$, there exist finite graphs G, H such that $R^{(H,n)} \not\cong R^{(G,n)}$.*

We leave open the question of necessary and sufficient conditions for the existence of a finite graph K so that $G \prec_n K$ and $H \prec_n K$ for finite graphs G and H.

3   Constructing Infinite Graphs by ($\leq n$)-extensions

We turn our attention now the the study of $R^{(G, \leq n)}$. Using the tools and methods of the previous section, we show that the behaviour of $R^{(G, \leq n)}$ closely mimics that of $R^{(G,n)}$. We also show that for any pair of finite graphs G and H, we have $R^{(G, \leq n)} \not\cong R^{(H,n)}$. This suggests that infinite graphs of the form $R^{(G, \leq n)}$

represent a new class of infinite n-ordered graphs that are strongly n-existentially closed.

We begin with preliminary results for $R^{(H,\leq n)}$. The proofs of these results follow similarly to work in [9] and in the previous section.

Theorem 8. *Let H be a finite non-trivial graph. The graph $R^{(H,\leq n)}$ is strongly n-e.c. but not strongly $n+1$-e.c..*

Corollary 4. *Let G be a strongly n-e.c. graph. If there exists a finite graph H such that $H \prec_{\leq n} G$ then G is not strongly $n+1$.e.c..*

Theorem 9. *Let G and H be finite graphs. If $H \leq G$, then $R^{(H,\leq n)} \leq R^{(G,\leq n)}$.*

Corollary 5. *For every non-trivial finite graph H, we have that every n-ordered graph is an induced subgraph of $R^{(H,\leq n)}$.*

Recall that the n-core of a countable graph Γ is the unique (up to isomorphism) maximum finite induced subgraph of minimum degree $n+1$ contained in Γ. If Γ has no such subgraph, then the n-core of Γ is defined to be K_1. In this case we say that Γ has a *trivial n-core*. Notice that a countable graph Γ is n-ordered if and only if Γ has a trivial n-core. Observe that if K is the smallest induced subgraph of Γ such that $K \prec_{\leq n} \Gamma$, then K is the n-core of Γ. The concept of n-core plays an important role in the study of infinite graphs of the form $R^{(H,\leq n)}$.

Theorem 10. *The finite graph H has a trivial n-core if and only if $R^{(H,\leq n)}$ is n-ordered.*

Corollary 6. *Every subgraph of $R^{(H,\leq n)}$ is n-ordered if and only if H has a trivial n-core.*

To find necessary and sufficient conditions such that $R^{(H,\leq n)} \cong R^{(G,\leq n)}$ we require results similar to Lemma 1 and Theorem 6.

Lemma 2. *Let G be a graph. Let $S \subseteq V(G)$ such that $|S| \leq n$. If G' is the graph formed from G by adding a single vertex x adjacent to each vertex in S, then $R^{(G,\leq n)} \cong R^{(G',\leq n)}$.*

Theorem 11. *If H and G are finite graphs so that $H \prec_{\leq n} G$, then $R^{(H,\leq n)} \cong R^{(G,\leq n)}$.*

Observe that if G is a non-trivial n-core, then the n-core of $R^{(G,\leq n)}$ is exactly G. To see this, consider any induced subgraph of $R^{(G,\leq n)}$. If such a subgraph contains a vertex not contained in a G, then such a subgraph necessarily has a vertex of degree at most n. Such a vertex can be found by considering the vertex with the largest index in such a subgraph (with respect to a natural enumeration of $R^{(G,\leq n)}$). This fact gives rise to the following result.

Theorem 12. *Finite graphs G and H have the same n-core if and only if $R^{(G,\leq n)} \cong R^{(H,\leq n)}$.*

Proof. Assume G and H have the same n-core, K. Observe $K \prec_{\leq n} G$ and $K \prec_{\leq n} H$. It follows from Corollary 11 that $R^{(K,\leq n)} \cong R^{(G,\leq n)}$ and $R^{(K,\leq n)} \cong R^{(H,\leq n)}$. The result follows by transitivity of isomorphism.

Assume now that G and H do not have the same n-core. Let G' be the n-core of G and H' be the n-core of H. By Theorem 11 it suffices to show $R^{(G',\leq n)} \not\cong R^{(H',\leq n)}$. Assume an isomorphism, $\phi : R^{(G',\leq n)} \to R^{(H',\leq n)}$ exists. The existence of such an isomorphism implies $R^{(G',\leq n)}$ and $R^{(H',\leq n)}$ have the same n-core, as the n-core of a countable graph is unique up to isomorphism. Since G' is the n-core of $R^{(G,\leq n)}$ and since H' is the n-core of $R^{(H,\leq n)}$ we have directly that $G' \cong H'$, a contradiction. □

Corollary 7. *If G and H are finite n-ordered graphs, then $R^{(G,\leq n)} \cong R^{(H,\leq n)}$.*

Theorem 12 gives a full classification of infinite graphs of the form $R^{(G,\leq n)}$. Every unique n-core, that is, every graph G with minimum degree $n+1$, generates a unique infinite graph $R^{(G,n)}$.

The characterizations and methods above can be used to show the class of the form $R^{(H,\leq n)}$ represent a class of infinite graphs distinct from those studied in [9].

Theorem 13. *If H and G are finite graphs then $R^{(H,n)} \not\cong R^{(G,\leq n)}$.*

The result in Theorem 5 closely mirrors one in [9] for $R^{(n)}$. However, by Theorem 13 we see the class of infinite graphs of the form $R^{(H,n)}$ where H is finite and n-ordered are not the only countable graphs for which every induced subgraph is n-ordered and every n-ordered graph appears as a subgraph. Classifying these universal countable infinite graphs is an interesting open area for research.

We conclude our study of $R^{(H,\leq n)}$ with a probabilistic construction generalizing the probabilistic construction of $R^{(n)}$ in [9]. The construction in [9] iteratively constructs an n-ordered graph so that at each step a single set of cardinality n is extended. In the construction below we modify this process so that at each step a collection of subsets $\{S_i\}$, $1 \leq i \leq n$, where $|S_i| = i$, are each extended.

Consider the following process which we name *Model* $(\leq n)$. Let $n \geq 1$ and let H_0' be a non-trivial finite graph, with vertex set $u_1^0, u_2^0, \ldots, u_{n_H}^0$. We form H_t' from H_{t-1}' by adding new vertices $\{x_1^t, x_2^t, \ldots, x_{n(t)}^t\}$ where $n(t) = \min\{n, V(H_{t-1}')\}$. We add these vertices so that x_i^t is adjacent to exactly i vertices of H_{t-1}' ($1 \leq i \leq n(t)$), and each vertex of x_k^j ($0 \leq j \leq t-1, 1 \leq k \leq n(j)$) of H_{t-1}' is adjacent to x_i^t with a probability exponentially proportional to j. Formally, for fixed $i \geq 1$, and each $S = \{x_{k_1}^{j_1}, x_{k_2}^{j_2}, \ldots, x_{k_i}^{j_i}\} \subset V(H_{t-1}')$, let $\mu(S) = 2^{-(j_1+j_2+\cdots+j_i)}$. Let

$$C_i^t = \sum_{S \subseteq V(H_{t-1}),|S|=i} \mu(S)$$

and let

$$\mathbb{P}(N_{H_t}(x_i^t) = S) = \frac{\mu(S)}{C_i^t}.$$

That is, for each x_i^t we choose $S_i \subset V(H'_{t-1})$ so that $|S| = i$. To form H'_{t-1} we add vertices $\{x_1^t, x_2^t, \ldots, x_{n(t)}^t\}$ and edges from x_i^t to all vertices in S for all $1 \le i \le n(t)$. A particular subset $S_i \subset V(H'_{t-1})$ is chosen with probability $\frac{\mu(S_i)}{C_i^t}$.

Theorem 14. *Let $H' = \lim_{t \to \infty} H'_t$, where H'_t was generated with Model ($\le n$). With probability 1, H' is strongly n-e.c..*

Theorem 15. *Let $H_0 = H'_0$ be a finite non-trivial graph. Let $H' = \lim_{t \to \infty} H'_t$, where H'_t was generated with Model ($\le n$). With probability 1, $H' \cong R^{(H_0, n)}$.*

The proof of Theorem 15 uses the same general method of *back and forth* construction of partial isomorphisms used in the proof of Lemma 1. This proof directly applies the results of Theorems 7 and 14.

We note that the analogous probabilistic construction of $R^{(G,n)}$, named *Model (n)* appears in [9] for the case $G = K_1$. The proofs of the following theorems follow directly from corresponding results in [9] for $R^{(n)}$.

Theorem 16. *Let $H' = \lim_{t \to \infty} H'_t$, where H'_t was generated with Model (n). With probability 1, H' is strongly n-e.c..*

Theorem 17. *Let $H_0 = H'_0$ be a finite non-trivial graph. Let $H' = \lim_{t \to \infty} H'_t$, where H'_t was generated with Model (n). With probability 1, $H' \cong R^{(H_0, n)}$.*

4   An Application of the Strong e.c. Property to Graph Distinguishing

The e.c. property allows us to exactly compare subsets of vertices to see how their roles within a network are similar and dissimilar. In this section we exploit the strong e.c property to show that for a class of infinite graphs one may distinguish similar vertices from one another by their relationship to independent sets.

Let G be a graph and $c : V(G) \to \{1, 2, 3, \ldots, k\}$. Let $\mathrm{aut}_c(G)$ be the set of automorphisms f of G with the property that for all $v \in V(G)$ we have $c(f(v)) = c(v)$. That is, $\mathrm{aut}_c(G)$ is the set of automorphisms of G that preserve the mapping c. We say that c is k-*distinguishing* if $\mathrm{aut}_c(G) = \{\epsilon\}$, where ϵ denotes the trivial automorphism. In other words, a labelling is distinguishing if it "breaks" all of the non-trivial automorphisms of G. The *distinguishing number of G*, denoted $D(G)$, is the least integer k such that G has a k-distinguishing labelling.

Distinguishing was first introduced by Albertson and Collins [3]. Since then authors have considered the distinguishing number of graph families [4,13], the complexity of the associated decision problems [17,23], generalizations with chromatic number [12,14], and even the distinguishing number of infinite graphs [19]. Of particular note is a conjecture regarding the distinguishing number of primitive countable homogeneous relational structures. In [20] the authors generalize distinguishing number for finite and countable homogeneous structures. They show that in most cases the distinguishing number is either 2 or infinite. They conjecture in fact that, other than a small number of exceptions, all primitive countable homogeneous relational structures with finite distinguishing number have distinguishing number 2.

In this section we introduce a new distinguishing distinguishing parameter – independent distinguishing. We call a k-distinguishing labelling *independent* if the pre-image of some label t, $1 \leq t \leq k$ is non-empty and induces an independent set in G. Let $D_i(G)$ be the independent distinguishing number of G.

We provide a sketch of the proof of the following result.

Theorem 18. *If G is strongly 1-e.c. and has countably many vertices, then $D_i(G) = 2$.*

We modify the technique used to show that the distinguishing number of the Rado graph is 2 [19]. We require the following notation: for G a countable graph, k a positive integer, $c : V(G) \to \{1, 2, \ldots, k\}$ and $1 \leq t \leq k$, let $G_c[t]$ be the set of vertices v of G so that $c(v) = t$. The key ingredients in the proof are the following two lemmas.

Lemma 3. *If G is strongly 1-e.c., then for all $x \in V(G)$, all finite $Y \subset V(G) \setminus x$ and for all $k > 0$ there exists an independent set $I_k \subset V(G) \setminus (\{x\} \cup Y)$ of cardinality k such that each vertex of I_k is correctly joined to x and Y.*

Proof. Let G be a strongly 1-e.c., graph. Consider $x \in V(G)$ and some finite $Y_0 \subset V(G) \setminus x$. Since G is strongly 1-e.c. there exists $z_1 \in V(G) \setminus (\{x\} \cup Y_0)$ such that z_1 is correctly joined to x and Y_0. Let $Y_1 = Y_0 \cup \{z_1\}$. Since G is strongly 1-e.c. there exists $z_2 \in V(G) \setminus (\{x\} \cup Y_1)$ such that z_2 is correctly joined to x and Y_1. Proceeding in this manner constructs the requisite independent set $z_1, z_2, \ldots, z_k$ for any $k > 0$. □

Lemma 4. *Let G be a strongly 1-e.c. infinite graph and $c : V(G) \to \{1, 2\}$. If*

1. *$G_c[2]$ is an independent set; and*
2. *each $v \in G_c[1]$ has a unique number of neighbours in $G_c[2]$,*

then c is independent 2-distinguishing labelling.

Proof. Let G be a strongly 1-e.c. infinite graph and let $c : V(G) \to \{1, 2\}$ satisfy 1 and 2. Consider $\phi \in \mathrm{aut}_c(G)$. By 2, we have that $\phi(u) = u$ for each $u \in G_c[1]$, as otherwise there exists $w \in G_c[i]$ such that $c(\phi(w)) \neq i$. Assume there exists $v_1 \neq v_2 \in V(G)$ so that $\phi(v_1) = v_2$. By the previous argument we have $v_1 \in G_c[2]$. Since G is strongly 1-e.c. there exists $z \in V(G) \setminus \{v_1, v_2\}$ that is correctly joined to v_1 and v_2. Since $G_c[2]$ is an independent set, we have that $z \in G_c[1]$. Since ϕ is an automorphism we have $zv_1 \in E(G)$ implies $\phi(z)\phi(v_1) = zv_2 \in V(G)$. This contradicts that z is correctly joined to v_1 and v_2. Thus no such z exists and we see that $\phi(v) = v$ for each $v \in G_c[2]$. Therefore $\mathrm{aut}_c(G)$ contains only the identity automorphism. □

We proceed now to outline the proof of Theorem 18. Let G be strongly 1-e.c. with countably many vertices. Let $v_1, v_2, \ldots$ be an enumeration of the vertices. Let $n_1, n_2, \ldots$ be a sequence of positive integers such that $\sum_{t=1}^{k} n_t < n_{k+1}$ for all $k > 1$. We construct a sequence of mappings $c_i : V(G) \to \{1, 2, 3\}$ such that for all $i \geq 1$ we have

- if $i' \leq i$ and $c_i(v_{i'}) = 1$, then $v_{i'}$ has exactly $n_{i'}$ neighbours in $G_{c_i}[2]$; and
- $G_{c_i}[2]$ is an independent set.

Let $c_0(v) = 3$ for all $v \in V(G)$. Proceeding inductively, we construct c_{i+1} from c_i as follows: If $c_i(v_{i+1}) = 2$, then we let $c_{i+1} = c_i$. Otherwise we let $c_{i+1}(v_{i+1}) = 1$. By construction, v_{i+1} has fewer than n_{i+1} neighbours x such that $c_i(x) = 2$. By applying Lemma 3, we can construct an independent set I such that $c_i(y) = 3$ for all $y \in I$. Relabeling each vertex in I to have label 2 will give v_{i+1} exactly n_{i+1} neighbours with label 2 with respect to the labelling c_i. This relabelling defines c_{i+1}.

Let $Y_i = G_{c_i}[2] \cup G_{c_i}[1]$ for all $i \geq 1$. Notice now that as $n \to \infty$ we have $|Y_i| \to \infty$ as $v_i \in Y_i$ for all $i \geq 1$. Taking the limit of this sequence of labellings yields a labelling in which no vertex is assigned label 3. To complete the proof one must verify that such a labelling satisfies the hypothesis of Lemma 4.

Corollary 8. *Let G be a finite graph and let $n \geq 1$. The following graphs have independent distinguishing number 2: R, $R^{(G,n)}$ and $R^{(G,\leq n)}$.*

5   Conclusion

Throughout our construction of $R^{(G,n)}$ and $R^{(H,\leq n)}$ we have assumed that both G and H are finite and simple. Though much of the analysis would remain unchanged if we drop the assumption of finiteness, proofs of theorems that depend on the finiteness of G and H no longer hold. As both $\prec_n$ and $\prec_{\leq n}$ remain transitive for infinite graphs, the study of $R^{(G,n)}$ and $R^{(H,\leq n)}$ when G and H are countably infinite may yield further insight in to graphs that are strongly n-e.c.. To consider graphs with parallel edges we can consider modifying the construction of $R^{(G,n)}$ to allow S to be a multi-set. In this case it can seen that the simple graph underlying $R^{(G,n)}$ is precisely $R^{(G,\leq n)}$.

In the context of online graph construction models, this limits of this work are clear – in the real world graphs are finite; there are no strongly n-e.c. graphs. It is possible that the graphs generated after finitely many steps by Model n and Model $(\leq n)$ have adjacency properties similar to n-e.c. property and strong n-e.c. property. Future work in this area should consider finite variants of Model n and Model $(\leq n)$, as well as how the finite versions of these models compare with standard online random graph models. As a first comparison, one might compare these models to *Degree Preferential Attachment* models. In a sense both Model $(\leq n)$ and Model (n) are preferential attachment models: as we proceed older vertices have higher degree than younger vertices and these higher degree vertices are more likely to be chosen for attachment in subsequent rounds. However in this case the preference for a vertex to be chosen is caused by its age rather than its degree. Regardless, it seems possible that the finite versions of these constructions will produce scale-free and possibly small world graphs.

References

1. Abbasi, A., Hossain, L., Leydesdorff, L.: Betweenness centrality as a driver of preferential attachment in the evolution of research collaboration networks. J. Inf. **6**(3), 403–412 (2012)
2. Albert, R., Barabási, A.L.: Statistical mechanics of complex networks. Rev. Mod. Phys. **74**(1), 47 (2002)
3. Albertson, M.O., Collins, K.L.: Symmetry breaking in graphs. Electron. J. Comb. **3**(1), 18 (1996)
4. Balachandran, N., Padinhatteeri, S.: Distinguishing chromatic number of random Cayley graphs. Discrete Math. **340**(10), 2447–2455 (2017)
5. Barabási, A.L., Albert, R.: Emergence of scaling in random networks. Science **286**(5439), 509–512 (1999)
6. Barabási, A.L., Bonabeau, E.: Scale-free networks. Sci. Am. **288**(5), 60–69 (2003)
7. Benzi, M., Klymko, C.: On the limiting behavior of parameter-dependent network centrality measures. SIAM J. Matrix Anal. Appl. **36**(2), 686–706 (2015)
8. Bonato, A.: The search for n-e.c. graphs. Contrib. Discret. Math. **4**(1), 40–53 (2009)
9. Bonato, A., Janssen, J., Wang, C.: The n-ordered graphs: a new graph class. J. Graph Theory **60**(3), 204–218 (2009)
10. Bondy, J.A., Murty, U.S.R.: Graph Theory. Springer, London (2008). https://doi.org/10.1007/978-1-84628-970-5
11. Cameron, P.J.: The random graph revisited. Eur. Congr. Math. **1**, 267–274 (2000)
12. Cheng, C.T.: On computing the distinguishing and distinguishing chromatic numbers of interval graphs and other results. Discrete Math. **309**(16), 5169–5182 (2009)
13. Choi, J.O., Hartke, S.G., Kaul, H.: Distinguishing chromatic number of Cartesian products of graphs. SIAM J. Discret. Math. **24**(1), 82–100 (2010)
14. Collins, K.L., Trenk, A.N.: The distinguishing chromatic number. Electron. J. Comb. **13**(1), 16 (2006)
15. Deijfen, M., Van Den Esker, H., Van Der Hofstad, R., Hooghiemstra, G.: A preferential attachment model with random initial degrees. Arkiv för Matematik **47**(1), 41–72 (2009)
16. Erdős, P., Rényi, A.: Asymmetric graphs. Acta Math. Hung. **14**(3–4), 295–315 (1963)
17. Eschen, E.M., Hoàng, C.T., Sritharan, R., Stewart, L.: On the complexity of deciding whether the distinguishing chromatic number of a graph is at most two. Discrete Math. **311**(6), 431–434 (2011)
18. Flaxman, A.D., Frieze, A.M., Vera, J.: A geometric preferential attachment model of networks. Internet Math. **3**(2), 187–205 (2006)
19. Imrich, W., Klavžar, S., Trofimov, V.: Distinguishing infinite graphs. Electron. J. Comb. **14**(1), R36 (2007)
20. Laflamme, C., Sauer, N., et al.: Distinguishing number of countable homogeneous relational structures. Electron. J. Comb. **17**(1), R20 (2010)
21. De Solla Price, D.: A general theory of bibliometric and other cumulative advantage processes. J. Am. Soc. Inf. Sci. **27**(5), 292–306 (1976)

22. Ravasz, E., Barabási, A.L.: Hierarchical organization in complex networks. Phys. Rev. E **67**(2), 026112 (2003)
23. Russell, A., Sundaram, R.: A note on the asymptotics and computational complexity of graph distinguishability. Electron. J. Comb. **5**(1), 23 (1998)
24. Telesford, Q.K., Joyce, K.E., Hayasaka, S., Burdette, J.H., Laurienti, P.J.: The ubiquity of small-world networks. Brain Connect. **1**(5), 367–375 (2011)
25. Wang, Z., Scaglione, A., Thomas, R.J.: Generating statistically correct random topologies for testing smart grid communication and control networks. IEEE Trans. Smart Grid **1**(1), 28–39 (2010)
26. Watts, D.J., Strogatz, S.H.: Collective dynamics of 'small-world' networks. Nature **393**(6684), 440 (1998)

The Robot Crawler Model on Complete k-Partite and Erdős-Rényi Random Graphs

A. Davidson[(✉)] and A. Ganesh

School of Mathematics, University of Bristol, University Walk, Bristol BS8 1TW, UK
angus.davidson@cantab.net, a.ganesh@bristol.ac.uk

Abstract. Web crawlers are used by internet search engines to gather information about the web graph. In this paper we investigate a simple process which models such software by walking around the vertices of a graph. Once initial random vertex weights have been assigned, the robot crawler traverses the graph deterministically following a greedy algorithm, always visiting the neighbour of least weight and then updating this weight to be the highest overall. We consider the maximum, minimum and average number of steps taken by the crawler to visit every vertex of firstly, sparse Erdős-Rényi random graphs and secondly, complete k-partite graphs. Our work is closely related to a paper of Bonato et al. who introduced the model.

MSC2010 Subject Classification. 60C05 · 05C80 · 05C81 · 05C85 · 90B15

1 Introduction

Using an analogy introduced by Messinger and Nowakowski [8], heuristically the robot crawler model can be viewed as a robot cleaning the nodes of a graph according to a greedy algorithm. Upon arriving at a given vertex the robot "cleans" the vertex, and then moves to its "dirtiest" neighbour to continue the process. Crawlers are of practical use in gathering information used by internet search engines, [4,7,10]. This particular version of the model was introduced by Bonato et al. [3] and we direct the reader to their paper for further insight into the problem's motivation and previous work done. There they considered the robot crawler performed on trees, complete k-partite graphs (with equal sized vertex classes), Erdős-Rényi random graphs and the preferential attachment model. The purpose of this paper is to offer an answer to open problems 1 and 2 posed there which relate to generalising their work concerning complete k-partite graphs and Erdős-Rényi random graphs.

The model introduced by Messinger and Nowakowski [8] is analogous to the robot crawler model, but the robot cleans edges, (which are weighted), rather than vertices. Models similar to those studied by Messinger and Nowakowski [8] were investigated by Berenbrink, Cooper and Friedetzky [1] and Orenshtein

© Springer Nature Switzerland AG 2019
K. Avrachenkov et al. (Eds.): WAW 2019, LNCS 11631, pp. 57–70, 2019.
https://doi.org/10.1007/978-3-030-25070-6_5

and Shinkar [11] who considered a class of random walks on graphs which prefer unused edges, although in their models the walker chooses independently among adjacent edges when they have all previously been traversed.

Given a finite connected undirected simple graph $G = G(V, E)$ we fix from outset an initial weighting; a bijective function $w_0 : V \to \{-n, -n + 1..., -1\}$ indicating the initial ranking of how dirty the vertices are. Here and henceforth "dirtiest"/"cleanest" refers to the vertex with the lowest/highest weight in a given set. At time 1 the robot visits the "dirtiest" node in V, i.e. $w_0^{-1}(-n)$. At time $t \in \mathbb{N}$ the robot updates the weight of the vertex visited to t. So if the robot visits vertex v at time t then $w_t(v) = t$ and $w_t(v') = w_{t-1}(v') \, \forall v' \in V, v' \neq v, t \in \mathbb{N}$. If all vertices then have positive weight, i.e. $\min_{y \in V}(w_t(y)) > 0$ then the algorithm terminates and we output $\mathcal{RC}(G, w_0) = t$; the number of steps taken to clean all vertices. Otherwise at time $t + 1$ the robot moves to vertex $argmin\{w_t(u) : (u, v) \in E\}$ i.e. the dirtiest neighbour of v at time t, and the process continues. As proved in [3], this algorithm will always terminate after a finite number of steps.

Using Ω_n to denote the set of $(n!)$ initial weightings we define

$$\mathrm{rc}(G) = \min_{w_0 \in \Omega_n} (\mathcal{RC}(G, w_0)) \text{ and}$$

$$\mathrm{RC}(G) = \max_{w_0 \in \Omega_n} (\mathcal{RC}(G, w_0)),$$

the minimum and maximum number of steps needed to clean all vertices of G.

Now supposing $\overline{w_0}$ is a uniformly chosen element of Ω_n we define the average number of steps needed to clean all vertices of G; $\overline{\mathrm{rc}}(G) = \mathbb{E}(\mathcal{RC}(G, \overline{w_0}))$.

We first consider the robot crawler number of sparse Erdős-Rényi random graphs in Sect. 2 before moving on to some results concerning the robot crawler on complete k-partite graphs in Sect. 3.

2 Erdős-Rényi Random Graph

We turn our attention to the main result of this paper: open problem 2 in [3]. In their paper Bonato et al. considered the robot crawler performed on $G(n, p)$ with $np \geq \sqrt{n \log n}$. We will prove the 2 results in Theorem 1 below which are similar to Corollary 2 and Theorem 8 in their work, but for much sparser graphs:

Theorem 1. *Let* $p = f(n) \log n / n$ *for some non-decreasing function* $f > 28$.

(i) $RC(G(n, p)) \leq n^{2+o(1)}$ *a.a.s.*
(ii) *W.l.o.g. for* $1 \leq i \leq n$ *fix* $w_0(v_i) = -i$. *Then*

$$\frac{\mathcal{RC}(G(n, p), w_0)}{\left(n + \frac{n}{f(n)}\right)} \xrightarrow{p} 1 \text{ as } n \to \infty$$

In particular we note that if $f(n) \to \infty$ as $n \to \infty$, however slowly, then $\frac{\mathcal{RC}(G(n,p), w_0)}{n} \xrightarrow{p} 1$ as $n \to \infty$.

Proof (of Theorem 1 (i)). We will use Lemma 1(5) from [3] which states that for any graph G, $RC(G) \leq n(\Delta + 1)^d$ where Δ is the maximum degree of a vertex in G, and d is the diameter of G.

The number of neighbours of v, a typical vertex of $G(n,p)$, is distributed $Bin(n-1,p)$. Hence,

$$\mathbb{P}(v \text{ has } \geq 2np \text{ neighbours}) = \mathbb{P}(Bin(n-1,p) \geq 2np)$$
$$= (1+O(1))\mathbb{P}(\mathcal{N}((n-1)p, (n-1)p(1-p)) \geq 2np)$$
$$\leq (1+O(1))\Phi\left(\frac{-np}{\sqrt{(n-1)p(1-p)}}\right)$$
$$\leq (1+O(1))\Phi(-\sqrt{np})$$
$$\leq (1+O(1))\frac{e^{-np/2}}{\sqrt{2\pi np}}$$
$$\leq (1+O(1))n^{-f(n)/2} \leq (1+O(1))n^{-14}$$

Hence by the union bound, $\mathbb{P}(\Delta \geq 2np) \leq (1+O(1))n^{-13}$

In a 2004 paper [5], (which extends the work of Bollobás [2]), Chung and Lu showed that a.a.s., $d = (1+o(1))\frac{\log n}{\log(np)}$ for $np \to \infty$. Putting these bounds together, a.a.s;

$$n(\Delta+1)^d \leq n(2np)^{(1+o(1))\frac{\log n}{\log np}}$$
$$= n\exp\left((1+o(1))\frac{\log n}{\log np}\log 2np\right)$$
$$= n^{2+o(1)}$$

To prove part (ii), we will have use for the following lemma:

Lemma 1. *Let* $Y = \sum_{i=1}^{n/7} X_i$ *where* $X_i \sim Geom(1 - (1-p)^i)$ *independently for each* $1 \leq i \leq n/7$. *For all* $\varepsilon > 0$,

$$\mathbb{P}\left((1-\varepsilon)\left(\frac{n}{7} + \frac{n}{f(n)}\right) < Y < (1+\varepsilon)\left(\frac{n}{7} + \frac{n}{f(n)}\right)\right) \overset{n\to\infty}{\longrightarrow} 1$$

Proof. To prove the upper bound we will use the following stochastic domination:
For $Z_1 \sim Geom(q)$ and $Z_2 \sim Exp(-\log(1-q))$, $Z_1 \preceq 1 + Z_2$. That is $\mathbb{P}(Z_1 \geq x) \leq \mathbb{P}(Z_2 + 1 \geq x)$ for all $x \geq 0$, or Z_1 stochastically dominates $Z_2 + 1$.
Indeed, for $x \geq 1$, $\mathbb{P}(Z_2 + 1 \geq x) = e^{(x-1)\log 1-q} = (1-q)^{x-1}$.
Also, $\mathbb{P}(Z_1 \geq x) = (1-q)^{\lceil x-1 \rceil} \leq \mathbb{P}(Z_2 + 1 \geq x)$.
Defining $E_i \sim Exp(i)$ for $1 \leq i \leq \frac{n}{7}$,

$$Y \preceq \frac{1}{-\log(1-p)}\sum_{i=1}^{n/7} E_i + \frac{n}{7}$$

Hence,

$$\mathbb{P}\left(Y > (1+\varepsilon)\left(\frac{n}{7} + \frac{n}{f(n)}\right)\right) \leq \mathbb{P}\left(\frac{1}{-\log(1-p)}\sum_{i=1}^{n/7} E_i > \frac{(1+\varepsilon)n}{f(n)} + \frac{\varepsilon n}{7}\right)$$

$$\leq \mathbb{P}\left(\frac{1}{-\log(1-p)}\sum_{i=1}^{n/7} E_i > (1+\varepsilon)\frac{n}{f(n)}\right)$$

$$\leq \mathbb{P}\left(\sum_{i=1}^{n/7} E_i > (1+\varepsilon)\log n\right)$$

since $-\log(1-p) \geq p = \frac{f(n)\log n}{n}$. It is an elegant fact that $\sum_{i=1}^{n/7} E_i \sim \max_{1 \leq i \leq n/7}\{E_1^i\}$ where $E_1^i \sim Exp(1)$ i.i.d..

Indeed suppose M_i, $1 \leq i \leq n/7$ are the ordered values in the array $\{E_1^i\}_{1 \leq i \leq n/7}$. That is $M_1 = \min(\{E_1^i\}_{1 \leq i \leq n/7})$, $M_2 = \min(\{E_1^i\}_{1 \leq i \leq n/7} \setminus M_1)$ etc. Then $M_1 \sim E_{n/7}$ since M_1 is the minimum of $n/7$ $Exp(1)$ random variables. If $M_1 = E_1^j$ say, then $E_1^i - M_1 \sim Exp(1)$ for all $1 \leq i \leq n/7, i \neq j$ by the memoryless property of the Exponential distribution. Therefore $M_2 - M_1 = \min(\{E_1^i - M_1\}_{1 \leq i \leq n/7, i \neq j}) \sim E_{n/7-1}$ since $M_2 - M_1$ is the minimum of $n/7 - 1$ $Exp(1)$ random variables and so on. Hence if we further define $M_0 = 0$ then

$$\max_{1 \leq i \leq n/7}\{E_1^i\} = M_{n/7} = \sum_{i=1}^{n/7}(M_i - M_{i-1}) \sim \sum_{i=1}^{n/7} E_{n/7+1-i} = \sum_{i=1}^{n/7} E_i.$$

We apply the union bound to deduce

$$\mathbb{P}\left(Y > (1+\varepsilon)\left(\frac{n}{7} + \frac{n}{f(n)}\right)\right) \leq ne^{-(1+\varepsilon)\log n} = n^{-\varepsilon} \xrightarrow{n\to\infty} 0$$

In proving the lower bound, we will use an even simpler stochastic domination:

For $T_1^i \sim Geom(1-(1-p)^i)$ and $T_2^i \sim Geom(ip)$ with $i \geq 1, ip < 1$, $T_1^i \succeq T_2^i$. This follows from the simple inequality $1-(1-p)^i \leq ip$ which holds $\forall i \geq 1$. Let $T = \sum_{i=1}^{n/f(n)\log n} T_2^i$. We find

$$\mathbb{P}\left(Y < (1-\varepsilon)\left(\frac{n}{7} + \frac{n}{f(n)}\right)\right)$$

$$\leq \mathbb{P}\left(T + \sum_{i=1+n/f(n)\log n}^{n/7} 1 < (1-\varepsilon)\left(\frac{n}{7} + \frac{n}{f(n)}\right)\right)$$

$$\leq \mathbb{P}\left(T < \frac{n}{f(n)} - \frac{\varepsilon n}{7} + \frac{n}{f(n)\log n}\right)$$

We recognise the relation between T and the coupon collector problem, (see for example [9]). Now,

$$\mathbb{E}(T) = \sum_{i=1}^{n/f(n)\log n} \frac{1}{ip} = \frac{\log n - \log(f(n)\log n) + O(1)}{p}$$

$$= \frac{n}{f(n)} - \frac{\log(f(n)\log n) - O(1)}{f(n)\log n}$$

$$\text{Var}(T) = \sum_{i=1}^{n/f(n)\log n} \frac{1-ip}{(ip)^2} \leq \frac{1}{p^2}\sum_{i=1}^{\infty}\frac{1}{(i)^2} = \frac{\pi^2}{6p^2}$$

We use Chebyshev's inequality to conclude

$$\mathbb{P}\left(Y < (1-\varepsilon)\left(\frac{n}{7} + \frac{n}{f(n)}\right)\right)$$

$$\leq \mathbb{P}\left(T < \frac{n}{f(n)} - \frac{\varepsilon n}{7} + \frac{n}{f(n)\log n}\right)$$

$$\leq \mathbb{P}\left(|T - \mathbb{E}(T)| > \mathbb{E}(T) - \frac{n}{f(n)} + \frac{\varepsilon n}{7} - \frac{n}{f(n)\log n}\right)$$

$$\leq \mathbb{P}\left(|T - \mathbb{E}(T)| > \frac{\varepsilon n}{7} - \frac{n(\log(f(n)\log n) + O(1))}{f(n)\log n}\right)$$

$$\leq \mathbb{P}\left(|T - \mathbb{E}(T)| > \frac{\varepsilon n}{14}\right) \quad \text{(for large enough } n\text{)}$$

$$\leq \left(\frac{\varepsilon n}{14}\right)^{-2}\text{Var}(T) = \frac{196\pi^2}{6(\varepsilon f(n)\log n)^2} \xrightarrow{n\to\infty} 0$$

We will prove Theorem 1 (ii) by showing high probability lower/upper bounds on $\mathcal{RC}(G(n,p), w_0)$. This is achieved by showing that with high probability this robot crawler number dominates/is dominated by a particular sum of geometrics. We will then use Lemma 1 to reach the final conclusion.

Fix the order of the vertices of $G(n,p)$ by initial weighting before we realise the edges of the random graph. So w.l.o.g. $w_0(v_i) = -i \,\forall 1 \leq i \leq n$.

In what follows we describe an exploration of the graph by the crawler which is "minimalistic" in its exposure of edges: Initially the presence or absence of every edge is unknown and undiscovered. Given any particular position of the crawler (which starts at vertex v_n at time 1), it proceeds to "check" the presence or absence of potential edges one by one starting with the dirtiest potential neighbour (at time 1 this will be vertex v_{n-1}), and then the next dirtiest and so on until an edge is found. Only those potential edges which were "checked" in this search for an edge are uncovered in this step. For example, supposing initially the dirtiest neighbour of v_n is v_i, then once the crawler has moved to v_i in step 2, the absence of edges between v_n and v_j for all $n-1 \leq j \leq i+1$ will have been established, the presence of the edge between v_n and v_i will have been discovered and no other information regarding presence/absence of edges will have been uncovered.

Proof (of Theorem 1 (ii)).
Lower Bound

We begin by showing $\mathbb{P}\left(\mathcal{RC}(G(n,p), w_0) \le (1-\varepsilon)\left(n + \frac{f(n)}{n}\right)\right) \overset{n\to\infty}{\longrightarrow} 0$, the lower bound. The crawler begins at time 1 at vertex v_n, initially the dirtiest node. Suppose that the crawler is positioned at vertex v, and that there are i vertices yet to be visited.

- If this is the crawler's first visit to v, no information is known about the presence of potential edges between v and yet unvisited vertices, hence the probability that v is connected to an unvisited vertex is $1 - (1-p)^i$ independently of all previous steps of the algorithm.

Otherwise, suppose that w was the vertex visited immediately after the crawler was last at vertex v.

- If w had already been cleaned, then necessarily, it is cleaner than any yet unvisited vertex which implies there are no edges between v and yet uncleaned vertices.
- If w had not already been cleaned, there are no edges between v and any uncleaned vertices which were initially *dirtier* than w, but presence of edges between v and uncleaned vertices initially cleaner than w is independent of all previous steps of the algorithm.

In any case, the probability v is connected to an unvisited vertex is $1-(1-p)^j$ for some $0 \le j \le i$. Hence, independently of all previous steps of the process, the probability v is connected to an unvisited vertex is less than $1 - (1-p)^i$. This implies the number of steps needed before reaching the next yet uncleaned vertex dominates a $Geom(1 - (1-p)^i)$ random variable, and $\forall \varepsilon > 0$

$$\mathbb{P}\left(\mathcal{RC}(G(n,p), w_0) \le (1-\varepsilon)\left(n + \frac{f(n)}{n}\right)\right)$$

$$\le \mathbb{P}\left(\sum_{i=1}^{n-1} Geom(1 - (1-p)^i) \le (1-\varepsilon)\left(n + \frac{f(n)}{n}\right)\right)$$

$$\le \mathbb{P}\left(Y \le (1-\varepsilon)\left(\frac{n}{7} + \frac{f(n)}{n}\right)\right) \overset{n\to\infty}{\longrightarrow} 0$$

by Lemma 1. For $\frac{n}{7}+1 \le i \le n-1$ we use the simple fact that $Geom(1-(1-p)^i) \ge 1$ in the second inequality.

Upper Bound

It remains to show that $\forall \varepsilon > 0$

$$\mathbb{P}\left(\mathcal{RC}(G(n,p), w_0) \ge (1+\varepsilon)\left(n + \frac{f(n)}{n}\right)\right) \overset{n\to\infty}{\longrightarrow} 0.$$

As in [3] we will consider 3 different phases of the crawling process. The 3 phases are required to control the dependence of the latter stages of the process on earlier movements of the robot crawler. The verticies are divided into 3 groups: "sets 1 to 3" say where vertices in set 1 are dirtier than those in set 2, and vertices in set 2 are dirtier than those in set 3. Phase 1 is designed so that all vertices in set 1 are cleaned and only vertices in sets 1 and 2 are visited. Phase 2 is designed so that all vertices in set 2 are cleaned and only vertices in sets 2 and 3 are visited. By Phase 3 the construction ensures that nothing has been revealed about the existence or otherwise of edges between vertices in set 1 and vertices in set 3. During Phase 1 the crawler may reveal missing edges between pairs of vertices in set 1, and similarly missing edges between pairs of vertices in set 3 during Phase 2, however with high probability the number of missing edges can be bounded. During Phase 3 all vertices in set 3 are cleaned and only vertices in sets 1 and 3 are visited. The independence of the existence/absense of edges between sets 1 and 3 together with the bound on missing edges is enough to ensure (with high probability) that the crawler will visit every vertex in set 3 within a specified number of steps.

Phase 1: As with the lower bound, the process will start from v_n, initially the dirtiest node and proceed to clean vertices of the graph. This phase ends when either of the following occur:

(a) $\frac{4n}{7}$ vertices have been cleaned.
(b) The crawler is not adjacent to any of the $n/7$ dirtiest (and as yet uncleaned) vertices, which are necessarily contained in $\{v_i, \frac{2n}{7} < i \leq n\}$.

We define the jump number $J(v_i), (1 \leq i \leq n)$ of a vertex v_i as the number of times any cleaner node was visited (potentially including repeated visits to the same node) before v_i was first cleaned itself. Intuitively it is the number of potential edges connected to v_i which were explored before one was first found, since each occurrence of a cleaner vertex being chosen by the crawler before vertex v_i implies a missing edge between the crawler's position at that time and v_i. If Phase 1 ends due to (a), and also the condition "$J(v) \leq n/7$ for all vertices" at the end of Phase 1 holds we will say that property P1 holds.

During each step of the crawling process in Phase 1, potential edges between the crawler's position and dirty nodes have not yet been exposed in previous steps. At each step, event (b) occurs if $n/7$ unexplored edges are not present in $G(n, p)$. This occurs with probability at most

$$(1 - p)^{n/7} = (1 - f(n) \log n/n)^{n/7} \leq n^{-f(n)/7} = o(n^{-3})$$

Hence by the union bound, with probability $1 - o(n^{-2})$ Phase 1 ends due to (a).

Further, as argued above "$J(v_i) \geq n/7$" implies that the first $n/7$ unexplored potential edges to v_i were not present. Again this has probability at most $(1 - p)^{n/7} = o(n^{-3})$, and hence by another application of the union bound, property P1 holds with probability $1 - o(n^{-2})$.

An important point to note is that only edges between vertices in $\{v_i, \frac{2n}{7} < i \leq n\}$ have been explored. Crucially for Phase 3, property P1 implies that each vertex v cleaned in this phase has had at most $2n/7$ potential edges exposed by the crawler (at most $n/7$ before the v was cleaned since $J(v) \le n/7$ and at most $n/7$ in the step after v was cleaned since (b) does not occur).

Phase 2: We continue to clean vertices until any one of the following occurs:

(a) The crawler is not adjacent to any as yet uncleaned vertex.
(b) There are $n/7$ uncleaned vertices remaining in $G(n,p)$.

If Phase 2 ends due to (b), and all vertices in $\{v_i, \frac{2n}{7} < i \le n\}$ have been cleaned by the end of the phase, then we say property P2 holds. As in Phase 1, Phase 2 ends due to (a) at each step if (at least) $n/7$ unexplored edges are not present in $G(n,p)$. Again we can conclude using the union bound that Phase 2 ends due to (b) with probability $1 - o(n^{-2})$.

Suppose now that $\exists v \in \{v_i, \frac{2n}{7} < i \le n\}$ such that v has not been cleaned by the crawler by the end of Phase 2. This would imply that $J(v) \ge \frac{n}{7}$ which as previously calculated has probability $o(n^{-3})$. Using this observation we again use the union bound to deduce:

$$\mathbb{P}(\{\text{P2 holds}\}|\{\text{Phase 2 ends due to (b)}\} \cap \{\text{P1 holds}\})$$
$$= 1 - o(n^{-2})$$

Hence, summarising what has been done so far,

$$\mathbb{P}(\{\text{P1 holds}\} \cap \{\text{P2 holds}\}) = 1 - o(n^{-2})$$

Phase 3: During this phase the crawler will continue to visit yet uncleaned vertices of $G(n,p)$ as well as revisiting some of the vertices which were cleaned during Phase 1. These vertices will have the smallest weight at this stage. This phase ends when any of the following occur:

(a) The crawler is not adjacent to any yet uncleaned vertex nor to any vertex which was cleaned during Phase 1 and has not yet been revisited in Phase 3.
(b) The phase takes longer than $2n/7$ steps.
(c) All vertices are cleaned.

If Phase 3 ends due to (c) then we say property P3 holds. In the explanation that follows, we condition on the event that P1 and P2 hold.

At each step of this phase, in total there are at least $3n/7$ "target" vertices which are yet to be visited at all or were cleaned in Phase 1 and have yet to be revisited in this phase. The reason for this is there are $4n/7$ vertices cleaned in Phase 1, $n/7$ vertices yet to be visited at all and this phase takes at most $2n/7$ steps. If the crawler has just revisited a vertex cleaned in Phase 1, P1 implies at most $2n/7$ potential edges adjacent to the vertex will have been explored earlier in the process, so at least $\frac{3n}{7} - \frac{2n}{7} = \frac{n}{7}$ potential edges to "target" vertices are still unexplored. Otherwise, if the crawler has just visited a vertex for the first time in the process then all ($\ge 3n/7$) potential edges to "target" vertices are unexplored. This is because crucially: no edges between $\{v_i, \frac{2n}{7} < i \le n\}$ and $\{v_i, 1 \le i \le \frac{2n}{7}\}$ are explored in Phase 1; P2 implies the uncleaned vertices at the beginning of Phase 3 are contained within $\{v_i, 1 \le i \le \frac{2n}{7}\}$ and as in earlier phases, the presence of potential edges between any possible current location of the crawler and yet unvisited vertices is still undetermined, and independent of previous steps of the process. Once again, the union bound tells us the probability we have (at least) $n/7$ unexplored edges not present in $G(n,p)$ during one of these steps, and hence that Phase 3 ends due to (a), is $o(n^{-2})$.

We now argue that with probability $o(n^{-2})$ Phase 3 ends due to (b). This is essentially a repeat of the argument in Phase 2. If Phase 3 ends due to (b) then $\geq \frac{n}{7}$ vertices cleaned in Phase 1 will have been revisited during Phase 3. If $v \in \{v_i, 1 \leq i \leq \frac{2n}{7}\}$ is still uncleaned at the end of the phase, then $J(v) \geq \frac{n}{7}$, since all vertices cleaned in Phase 1 will be cleaner than v before it is itself cleaned. Once again, this has probability $o(n^{-3})$ and applying the union bound:

$$\mathbb{P}(\text{P3 holds}\}|\{\text{Phase 3 ends due to (b) or (c)}\} \cap \{\text{P2 holds}\} \cap \{\text{P1 holds}\})$$
$$= 1 - o(n^{-2})$$

We can now conclude that:

$$\mathbb{P}(\{\text{P3 holds}\}|\{\text{P1 holds}\} \cap \{\text{P2 holds}\}) = 1 - o(n^{-2})$$

and hence bringing together earlier calculations

$$\mathbb{P}(\{\text{P1, P2, P3 hold}\}) = 1 - o(n^{-2})$$

If $\widehat{Y} := (Y|Y \leq 2n/7)$ then conditional on P1, P2 and P3, Phases 1 and 2 will take $n - n/7$ steps and Phase 3 will take a number of steps distributed as $\widehat{Y}$. Indeed, during Phase 3 when there are x yet uncleaned vertices in $\{v_i, 1 \leq i \leq \frac{2n}{7}\}$, (and hence x unexplored edges from the crawler's current position and these vertices), the probability the crawler will be adjacent to at least one of them is given by $1 - (1-p)^x$. If the crawler continues to visit vertices with unexplored edges to all x yet uncleaned vertices then the probability the crawler will reach one of these x vertices in the next y steps is given by $\mathbb{P}(Geom(1 - (1-p)^x) \leq y)$. And so

$$\mathbb{P}\left(\mathcal{RC}(G(n,p), w_0) \geq (1+\varepsilon)\left(n + \frac{f(n)}{n}\right)\right)$$
$$\leq \mathbb{P}\left(\mathcal{RC}(G(n,p), w_0) \geq (1+\varepsilon)\left(n + \frac{f(n)}{n}\right) \Big| \{\text{P1, P2, P3 hold}\}\right)$$
$$+ \mathbb{P}\left(\{\text{P1, P2, P3 hold}\}^C\right)$$
$$\leq \mathbb{P}\left(\widehat{Y} + \frac{6n}{7} \geq (1+\varepsilon)\left(n + \frac{f(n)}{n}\right)\right) + o(n^{-2})$$
$$\leq \mathbb{P}\left(\widehat{Y} \geq (1+\varepsilon)\left(\frac{n}{7} + \frac{f(n)}{n}\right)\right) + o(n^{-2})$$
$$\leq \mathbb{P}\left(Y \geq (1+\varepsilon)\left(\frac{n}{7} + \frac{f(n)}{n}\right)\right) + o(n^{-2}) \xrightarrow{n \to \infty} 0$$

again, by Lemma 1.

3   Complete k-Partite Graphs

We now consider a number of results concerning the robot crawler on complete k-partite graphs.

3.1   Results

Given some constants $c_1 \geq c_2 ..., \geq c_k$, $\sum_{i=1}^{k} c_i = 1$, $k \geq 3$ consider the robot crawler model performed on the complete k-partite graph G_n induced by vertex sets $V_1, V_2, ..., V_k$ where $|V_i| = c_i n \; \forall 1 \leq i \leq k$.

Theorem 2.

(i) For $c_1 \leq \frac{1}{2}$, $rc(G_n) = n$
(ii) For $c_1 > \frac{1}{2}$, $rc(G_n) = 2nc_1 - 1$

Theorem 3.

(i) For $c_2 \leq \frac{1}{2}(1 - c_1)$, $RC(G_n) = n + c_1 n - 1$
(ii) For $c_2 > \frac{1}{2}(1 - c_1)$, $RC(G_n) = 2(n - c_2 n)$

Theorem 4.

(i) For $c_1 < \frac{1}{2}$, $\overline{rc}(G_n) = n + O(1)$
(ii) For $c_1 = \frac{1}{2}$, $\overline{rc}(G_n) = n + O(n^{\frac{1}{2}})$
(iii) For $c_1 > \frac{1}{2}$, $\overline{rc}(G_n) = 2nc_1 + O(1)$

In particular we note that for $c_1 \neq \frac{1}{2}$, $\overline{rc}(G_n) = rc(G_n) + O(1)$, which refines Theorem 6 in [3] if we take $G_n = K^k_{n/k}$, the complete k-partite graph induced by k vertex sets each of size $\frac{n}{k}$.

3.2   Proofs

We begin with the more straightforward proofs of Theorems 2 and 3 before approaching the more complex treatment of Theorem 4.

Proof (of Theorem 2). It is easy to construct a Hamiltonian path to verify part (i); for example, since the minimum vertex degree is at least $n/2$ we can apply Dirac's theorem on Hamiltonian cycles. With an appropriate choice of w_0 we can force the crawler to follow this Hamiltonian path which visits all nodes in n steps. For part (ii) we note that once the crawler is in set V_1 (which takes at least 1 step) it must return at least $c_1 n - 1$ times. Whenever the crawler is in set V_1 it will take at least 2 steps of the algorithm before the crawler returns since there are of course no edges between vertices in V_1. Hence $rc(G_n) \geq 1 + 2(nc_1 - 1)$. Noting that $|V_1| > |V \setminus V_1|$, the bound can be achieved if the crawler starts in V_1 and oscillates between V_1 and $V \setminus V_1$, e.g. if $w_0(v) < w_0(u) \; \forall v \in V_1, u \in V \setminus V_1$.

Given an initial weighting w_0, for $1 \leq i \leq k$ define the surplus of vertex set i $(=: S_{w_0}(i))$ to be the number of uncleaned vertices remaining in V_i at the moment all vertices in $V \setminus V_i$ have been cleaned. Clearly $S_{w_0}(i) = 0$ for all but one value of i. Further define $S_{w_0} = \sum_{i=1}^{k} S_{w_0}(i) = \max_{1 \leq i \leq k}(S_{w_0}(i))$. A crucial observation is that $\mathcal{RC}(G_n, w_0) = n + S_{w_0} - 1$. Indeed suppose $S_{w_0}(i) > 0$, then immediately after the time step $(t = n - S_{w_0}(i))$ when all vertices in $V \setminus V_i$ have been cleaned the crawler will alternate between V_i and $V \setminus V_i$ until all remaining $S_{w_0}(i)$ uncleaned vertices of V_i have been cleaned which will take a further $2S_{w_0}(i) - 1$ steps.

Proof (of Theorem 3). Clearly $S_{w_0} \leq \max_{1 \leq i \leq k} |V_i| = c_1 n$. Part (i) now amounts to showing that if $c_2 \leq \frac{1}{2}(1 - c_1)$ then $\exists w_0$ such that $S_{w_0} = c_1 n$. This follows in part since if $k \geq 4$ it is possible to clean $V \setminus V_1$ in $|V \setminus V_1|$ steps using Theorem 2 (i) on the complete $(k-1)$-partite graph induced by vertex sets $V_2, ..., V_k$, in which case $S_{w_0}(1) = c_1 n$. Finally, if $k = 3$ then necessarily $c_2 = c_3$ and again it is of course possible to clean $V \setminus V_1$ in $|V \setminus V_1|$ steps simply by alternating between V_2 and V_3 for the first $2c_2 n$ steps.

Suppose now $c_2 > \frac{1}{2}(1 - c_1)$. If $S_{w_0}(2) > 0$, then $S_{w_0}(1) = 0$. It takes at least $2nc_1 - 1$ steps to clean all vertices of V_1 at which point there are at most $n - 2nc_1 + 1$ vertices in V_2 not yet visited by the crawler. It will then take at most $2(n - 2nc_1 + 1) - 1$ steps to clean the remainder of the vertices, which implies $\mathrm{RC}(G_n) \leq 2(n - 2nc_1 + 1) - 1 + 2nc_1 - 1 = 2n(1 - c_1)$, less than the required upper bound.

If conversely $S_{w_0}(2) = 0$, then when V_2 is fully cleaned there are uncleaned vertices elsewhere in V. We first note that it takes at least $2nc_2 - 1$ steps to clean all vertices of V_2 at which point there are at most $n - 2nc_2 + 1$ vertices in V not yet visited by the crawler. From this point it will take at most $2(n - 2nc_2 + 1) - 1$ steps to clean the remainder of the vertices, which gives the required upper bound $\mathrm{RC}(G_n) \leq 2n(1 - c_2)$.

Consider $w_0 \in \Omega_n$ with set V_2 being the $|V_2|$ dirtiest, and V_1 the $|V_1|$ cleanest vertices of V. That is $\bigcup_{j=0}^{c_2 n - 1} w_0^{-1}(-n + j) = V_2$ and $\bigcup_{j=1}^{c_1 n} w_0^{-1}(-j) = V_1$, then the part (ii) bound is attained.

We now turn our attention to Theorem 4, the main result of the section. Let $m_i = \max(x : \exists y \geq 0 \text{ s.t. } y + x \text{ of the } 2y + x \text{ cleanest vertices lie in set } V_i)$. That is, $m_i(= m_i(w_0)) = \max(x : \exists y \geq 0 \text{ s.t. } |\bigcup_{j=1}^{2y+x} w_0^{-1}(-j) \cap V_i| = y + x)$. Stochastically m_i is the record of an n step simple random walk on $\mathbb{Z}$, conditioned to be at a fixed position at time n. This random walk starts at the origin at time 0 and jumps up (down) by 1 at time t if $w_0^{-1}(-t) \in V_i$ $(w_0^{-1}(-t) \in V \setminus V_i)$, and finishes at time n in position $|V_i| - |V \setminus V_i| = 2nc_i - n$. More on this shortly.

Lemma 2. $S_{w_0}(i) \leq m_i$.

Proof. W.l.o.g. take $i = 1$. First note that if $m_1 = |V_1|$ then $S_{w_0}(1) \leq m_1$ trivially. We now assume $m_1 < |V_1|$. Consider $x \in V_1$ defined to be the $(m_1 + 1)^{\text{st}}$ cleanest vertex in V_1, and suppose it is also the $(m_1 + 1 + t)^{\text{th}}$ cleanest vertex in V overall, (so $w_0(x) = -(m_1 + 1 + t)$). So, there are m_1 vertices cleaner than x

in V_1 and t cleaner than x in $V \setminus V_1$. Clearly $t \geq 1$ by the definition of m_1. Let v be the first vertex cleaned of the $(m_1 + 1 + t)$ cleanest of V. If $v = x$ then we are done. If $v \neq x$ then $v \in V \setminus V_1$ and v must have been cleaned immediately after some node $u \in V_1$ where $w_0(u) = -(m_1 + 1 + t + l)$ for some $l > 0$ and $\cup_{i=1}^{l} w_0^{-1}(-m_1 - 1 - t - i) \subset V_1$. By the definition of m_1, necessarily $l < t$, (and all other nodes must have already been cleaned by the crawler). It is clear how the crawler will then proceed, alternating between V_1 and $V \setminus V_1$ until x is cleaned at which point there will be $t - l > 0$ uncleaned vertices in $V \setminus V_1$, and hence $S_{w_0}(1) \leq m_1$.

It is not difficult to construct a graph with some initial vertex weights such that $S_{w_0}(1) < m_1$. As a simple example, consider the complete 3-partite graph G induced by V_1, V_2, V_3 with $|V_1| = 3, |V_2| = 3, |V_3| = 1$ and V_1 consisting of the 3 cleanest vertices of G. In this case $m_1 = 3$ but $S_{w_0}(1) \leq 2$.

We now make the link between m_1 and the record of a simple random walk bridge, noting the start and end points of this bridge can be different. Let $(Z(t))_{t \geq 0}$ be a random walk on $\mathbb{Z}$ starting from $Z(0) = 0$ with $p = \mathbb{P}(Z(t + 1) - Z(t) = 1)$ and $q = \mathbb{P}(Z(t + 1) - Z(t) = -1) = 1 - p \; \forall t \geq 0$. Then $(Z(t)|Z(n) = y)_{0 \leq t \leq n}$ is a random walk bridge. Necessarily, of the n steps of this random walk bridge, $\frac{n+y}{2}$ are up and $\frac{n-y}{2}$ are down. The random walk bridge can equivalently be defined as a uniformly random ordering of these up and down steps.

For $0 \leq t \leq n$ define $U(t) := |v \in V_1, w_0(v) \geq -t|$, the number of vertices in V_1 initially among the t cleanest of V, $D(t) := |v \in V \setminus V_1, w_0(v) \geq -t| = t - U(t)$ and $X(t) := U(t) - D(t) = 2U(t) - t$. Of course $U(n) = |V_1|$ and $D(n) = |V \setminus V_1|$, and $(X(t))_{0 \leq t \leq n}$ is generated by a uniformly random ordering of these $|V_1|$ up steps and $|V \setminus V_1|$ down steps. $(X(t))_{0 \leq t \leq n}$ is therefore a random walk bridge starting at $X(0) = 0$ and finishing at $X(n) = |V_1| - |V \setminus V_1|$. With $p = c_1$, $q = 1 - c_1$ and $y = |V_1| - |V \setminus V_1|$; $(X(t))_{0 \leq t \leq n} \sim (Z(t)|Z(n) = |V_1| - |V \setminus V_1|)_{0 \leq t \leq n}$. We could equally have defined $m_1 = \max_{0 \leq t \leq n} \{X(t)\}$.

Lemma 3. For $c_i < 0.5$, $\mathbb{E}(m_i) \leq \frac{2c_i}{1 - 2c_i}$.

Proof. Again, w.l.o.g. take $i = 1$. Let $h_j = \mathbb{P}(\max_{t \geq 0}(Z(t)) \geq j)$. As a simple consequence of the Markov property, for $j \geq 1$:

$$h_j = \mathbb{P}(Z(1) = 1)\mathbb{P}(\max_{t \geq 0}(Z(t)) \geq j | Z(1) = 1)$$
$$+ \mathbb{P}(Z(1) = -1)\mathbb{P}(\max_{t \geq 0}(Z(t)) \geq j | Z(1) = -1)$$
$$= c_1 h_{j-1} + (1 - c_1) h_{j+1}$$

Using $c_1 < \frac{1}{2}$ together with the initial condition $h_0 = 1$ we find that $h_j = (\frac{c_1}{1-c_1})^j \ \forall j \geq 0$. Now

$$\mathbb{P}(m_1 \geq j) = \mathbb{P}(\max_{0 \leq t \leq n} (X(t)) \geq j)$$

$$= \mathbb{P}(\max_{0 \leq t \leq n} (Z(t)) \geq j | Z(n) = |V_1| - |V \setminus V_1|)$$

$$\leq \mathbb{P}(\max_{0 \leq t \leq n} (Z(t)) \geq j | Z(n) \geq |V_1| - |V \setminus V_1|)$$

$$\leq \frac{\mathbb{P}(\max_{0 \leq t \leq n}(Z(t)) \geq j)}{\mathbb{P}(Z(n) \geq |V_1| - |V \setminus V_1|)}$$

$$\leq 2\mathbb{P}(\max_{0 \leq t \leq n} (Z(t)) \geq j)$$

The last inequality follows since $Z(n)$ is a Binomial random variable with mean and median $|V_1| - |V \setminus V_1|$. The first inequality follows from a simple coupling argument. If we are given a realisation of $(X(t))_{1 \leq t \leq n}$, and some integer $0 \leq C \leq |V_1|$, we can define the random path $(Z_1(t))_{1 \leq t \leq n}$ by taking C of the down steps of $X(t)$ chosen uniformly at random among all of the $\binom{|V_1|}{C}$ possibilities and flipping them to up steps. Clearly, $Z_1(t) \geq X(t)$ $\forall 1 \leq t \leq n$, and hence $\max_{0 \leq t \leq n}(Z_1(t)) \geq \max_{0 \leq t \leq n}(X(t))$. If we initially let $C \sim \frac{1}{2}(Z(n) - (|V_1| - |V \setminus V_1|))|Z(n) \geq |V_1| - |V \setminus V_1|$ then it is also clear that $Z_1(t) \sim Z(t) | Z(n) \geq |V_1| - |V \setminus V_1|$.

Concluding the argument:

$$\mathbb{E}(m_1) \leq 2 \sum_{j=1}^{\infty} \mathbb{P}(\max_{0 \leq t \leq n} (Z(t)) \geq j)$$

$$\leq 2 \sum_{j=1}^{\infty} \mathbb{P}(\max_{0 \leq t}(Z(t)) \geq j) = 2 \sum_{j=1}^{\infty} h_j = \frac{2c_1}{1 - 2c_1}$$

We can now conclude part (i) of Theorem 4. For $c_1 < 0.5$:

$$\overline{rc}(G_n, w_0) = \mathbb{E}(n + S_{w_0} - 1) \leq n + \mathbb{E}\left(\sum_{i=1}^{k} m_i\right) \leq n + \sum_{i=1}^{k} \frac{2c_i}{1 - 2c_i} = n + O(1)$$

In proving Lemma 3, we linked m_1 to the maximum of a Random Walk Bridge $X(t)$ with the property that $X(0) > X(n)$, and eventually used the expected maximum level reached by a Random Walk with negative drift. Using a similar strategy to conclude part (ii) of Theorem 4 where $c_1 > 0.5$ wouldn't work since of course, the expected maximum reached by a Random Walk with positive drift is unbounded. To navigate this problem we will reverse time on the Random Walk Bridge.

Corollary 1. *For $c_1 > \frac{1}{2}$, $\mathbb{E}(m_1) \leq 2nc_1 - n + \frac{2(1-c_1)}{2c_1-1}$*

Proof. For $1 \leq t \leq n$ define $\widehat{X}(t) = X(n-t)$. $\widehat{X}(t)$ is again a Random Walk Bridge, but with $\widehat{X}(0) = 2nc_1 - n$ and $\widehat{X}(n) = 0$. The key point here is that $\left(\widehat{X}(t)|c_1 = \alpha\right) \sim (2nc_1 - n + X(t)|c_1 = 1 - \alpha)$, so

$$\mathbb{E}(m_1) = \mathbb{E}\left(\max_{0 \leq t \leq n}\{X(t)\}\right) = \mathbb{E}\left(\max_{0 \leq t \leq n}\{\widehat{X}(t)\}\right) \leq 2nc_1 - n + \frac{2(1 - c_1)}{2c_1 - 1}$$

by Lemma 3.

We have now shown that for $c_1 > 0.5$, $\overline{\text{rc}}(G_n) \leq n + \mathbb{E}(m_1) \leq 2nc_1 + \frac{2(1-c_1)}{2c_1-1}$ which, together with part (ii) of Theorem 2, completes the proof of part (iii) of Theorem 4.

Finally, Godrèche et al. [6] prove that for $c_1 = 0.5$, $\mathbb{E}(\max_{0 \leq t \leq n}(X(t))) = \sqrt{\frac{\pi n}{8}}$. Part (ii) of Theorem 4 follows.

References

1. Berenbrink, P., Cooper, C., Friedetzky, T.: Random walks which prefer unvisited edges: exploring high girth even degree expanders in linear time. Random Struct. Algorithms **46**(1), 36–54 (2015)
2. Bollobás, B.: The diameter of random graphs. Trans. Am. Math. Soc. **267**(1), 41–52 (1981)
3. Bonato, A., del Río-Chanona, R.M., MacRury, C., Nicolaidis, J., Pérez-Giménez, X., Prałat, P., Ternovsky, K.: The robot Crawler number of a graph. In: Gleich, D.F., Komjáthy, J., Litvak, N. (eds.) WAW 2015. LNCS, vol. 9479, pp. 132–147. Springer, Cham (2015). https://doi.org/10.1007/978-3-319-26784-5_11
4. Brin, S., Page, L.: Anatomy of a large-scale hypertextual web search engine. In: 7th International World Wide Web Conference (1998)
5. Chung, F., Linyuan, L.: The diameter of sparse random graphs. Adv. Appl. Math. **26**(4), 257–279 (2001)
6. Godrèche, C., Majumdar, S.N., Schehr, G.: Record statistics for random walk bridges. J. Stat. Mech.: Theory Exp. **2015**(7), P07026 (2015)
7. Henzinger, M.R.: Algorithmic challenges in web search engines. Internet Math. **1**(1), 115–123 (2004)
8. Messinger, M.-E., Nowakowski, R.J.: The robot cleans up. In: Yang, B., Du, D.-Z., Wang, C.A. (eds.) COCOA 2008. LNCS, vol. 5165, pp. 309–318. Springer, Heidelberg (2008). https://doi.org/10.1007/978-3-540-85097-7_29
9. Mitzenmacher, M., Upfal, E.: Probability and Computing: Randomized Algorithms and Probabilistic Analysis. Cambridge University Press, Cambridge (2005)
10. Olston, C., Najork, M.: Web crawling. Found. Trends Inf. Retr. **4**(3), 175–246 (2010)
11. Orenshtein, T., Shinkar, I.: Greedy random walk. Comb. Probab. Comput. **23**(02), 269–289 (2014)

Estimating the Parameters
of the Waxman Random Graph

Matthew Roughan$^{(\boxtimes)}$, Jonathan Tuke, and Eric Parsonage

ARC Centre of Excellence for Mathematical and Statistical Frontiers,
University of Adelaide, Adelaide, Australia
{matthew.roughan,simon.tuke,eric.parsonage}@adelaide.edu.au

Abstract. The Waxman random graph is useful for modelling physical networks where the increased cost of longer links means they are less likely to be built, and thus less numerous. The model has been in continuous use for over three decades with many attempts to match parameters to real networks, but only a few cases where a formal estimator was used. Even then the performance of the estimator was not evaluated. This paper presents both the first evaluation of formal estimators for these graphs, and a new Maximum Likelihood Estimator with $O(e)$ computational complexity where e is the number of edges in the graph, and requiring only link lengths as input, as compared to all other algorithms which are $\Omega(n^2)$.

1 Introduction

The study of random graphs provides insight into the formation of real networks and allows synthesis of test networks for use in simulations. There are many alternative approaches, but within these random graphs explicitly incorporating underlying geometry became popular with the introduction of the Waxman random graph [1], proposed as an alternative to the Gilbert-Erdős-Rényi (GER) random graph [2,3] as a more realistic setting for testing networking algorithms. The Waxman graph has subsequently been used in applications as wide-ranging as computer networks, transportation and biology (Waxman's original paper is listed by Google Scholar as having been cited well over 3000 times).

The GER random graph links every pair of vertices independently with a fixed probability, whereas the Waxman graph reflects that in real networks longer links are often more costly or difficult to construct, and their existence therefore less likely. It links nodes i and j with a probability given by a function of their distance d_{ij}. The form chosen by Waxman was the exponential, *i.e.*,

$$p(d_{ij}) = q \exp\left(-sd_{ij}\right), \tag{1}$$

for parameters $s \geq 0$ and $q \in (0,1]$. A Waxman random graph is generated by randomly choosing a set of points, and then linking these independently according to the *distance deterrence* function (1).

Despite frequent use of the Waxman graph there is little formal literature on how to estimate the parameters (q, s) from a given graph, or set of graphs. In

© Springer Nature Switzerland AG 2019
K. Avrachenkov et al. (Eds.): WAW 2019, LNCS 11631, pp. 71–86, 2019.
https://doi.org/10.1007/978-3-030-25070-6_6

many works the parameter values have been chosen arbitrarily with the authors giving little justification for the values used. In other cases authors use the parameter values given in earlier works without regard for their applicability. A few works use more careful estimates, but do not evaluate estimator performance.

Within the statistical social network modelling community there has been considerable work on graph estimation, however, the focus of this work has only rarely touched on graphs with a spatial embedding. The work most closely connected to ours considers the case where the underlying space and distances are *latent* in the sense that they are not measured, and hence must be estimated as part of the problem. For instance see [4]. This is both a harder, and easier problem: harder algorithmically because one must estimate many more variables, but easier because the latent parameters are free to adjust to the data.

This paper presents a comparison of estimators of the parameters of the Waxman graph, including a new Maximum Likelihood Estimator (MLE). We demonstrate that its performance is close to the Cramér-Rao lower bound. The comparison shows its advantages in:

- accuracy;
- computational complexity (the MLE is $O(e)$ as compared to $\Omega(n^2)$ for alternatives); and
- reduced input (it uses only the lengths of observed links or just a sample of such links).

Finally we use the MLE to estimate the distance dependence of an Internet dataset. Additional datasets (and evaluations) are included in a pre-print of this work available at [5].

2   Background and Related Work

A graph (or network) consists of a set of n vertices (synonymously referred to as nodes) which without loss of generality we label $\mathcal{V} = \{1, 2, \ldots, n\}$, and a set of edges (or links) $\mathcal{E} \subset \mathcal{V} \times \mathcal{V}$. We denote the number of edges by $e = |\mathcal{E}|$. Here we are primarily concerned with undirected graphs.

The GER random graph $G_{n,p}$ of n vertices is constructed by assigning independently and with a fixed probability p each pair of nodes (i, j) to be in $\mathcal{E}$. The Waxman random graph [1] generalises this by making the probability that each pair of nodes is an edge dependent on the distance between the nodes. Some examples indicating the scope and durability of the model include [6–19].

The Waxman graph was originally defined using Euclidean distances on a rectangle or straight line with points on an integer grid, but most subsequent work has considered graphs defined with points randomly placed in the unit square. However, there is no impediment to choosing points in an arbitrary convex region with an arbitrary distance metric. Even convexity is not strictly required, except where there is the notion that the links are physical, and must themselves lie in the region of interest. Hence, we define a Waxman graph by placing n nodes uniformly at random within some defined region R of a metric space Ω with a distance metric $d(x, y)$ and each pair of nodes is made adjacent independently with probability given by (1).

The function (1) differs from Waxman's original parameterisation [1] $p(d_{ij}) = \beta \exp(-d_{ij}/L\alpha)$, where $\alpha \in (0, 1]$ and $\beta > 0$, and L is the greatest distance possible between any two points in the region of interest. Unfortunately, previous authors have confused this notation by reversing the roles of α and β with almost equal regularity, to the point where parameters chosen in one paper have been reversed in another purporting to compare results. Hence, we provide an alternate parameterisation (1) chosen with the estimation problem in mind, using a parameter s with units related to distance instead of the dimensionless α as more meaningful for real problems.

Properties such as connectivity and path lengths in Waxman graphs have been studied in several works [6, 20–22]. There are works on estimation of parameters of other random graph models, for instance see [23], but few have considered the estimation of the Waxman graph parameter. Indeed, some previous works compared graphs merely by visual inspection.

Despite the presence of the exponential function, the Waxman graph is not in the class of Exponential Random Graphs (ERG) [24, 25], where the exponential applies to the probability of a particular graph (not a particular link), as a function of the overall graph properties, such as the number of triangles. Hence the large literature on estimation of ERGs is not specifically applicable.

The first work we are aware of which attempted to estimate the parameters of a Waxman graph from real data is that of [26]. Using a large set of real Internet and geographic data the authors found that an exponential distance-based probability was reasonable for physical links between routers. The authors conducted the study with some care, comparing two sets of data and finding consistent results by using a log-linear regression of the link distance function against the link distances. However they made no effort to consider the efficiency or accuracy of their estimator, which we shall do here.

Other related work considers estimation of graph parameters of models where the nodes have some latent properties. The closest related work in that vein is where the latent properties of the nodes are distances [4], to be estimated as part of the overall problem. This is both a harder, and easier problem: harder because one must estimate many more variables, but easier because there are so many more degrees of freedom that almost any set of data should be possible to fit against such a model. In any case, these estimation techniques do not apply to these graphs, where the distances are not latent.

Estimation was used on the similar model [19] by determining the parameters of a binary Generalised Linear Model (GLM). We adapt this approach to the almost equivalent Waxman graphs, and compare it to alternatives: the method is accurate, but computationally complex both in time, and memory.

The underlying assumption of [19, 26] is that *all* of the distances are known, even for links that do not exist. This is possible if all node locations are known, but it is sometimes only possible to measure the length of a link that exists. Consider also graphs for which the "distance" is not a physical quantity, but rather a cost or an administratively configured link-weight, which is not defined except where a link exists. We present here a MLE that can work with only defined links.

Techniques that depend on existence tests of all edges have time complexity at least $O(n^2)$, where n is the number of nodes in the graph, whereas our MLE is $O(e)$, where e is the number of edges. This can result in a dramatic improvement in computation time as large real graphs are often very sparse, and we can achieve fast computation even for dense graphs by sampling using this approach.

Waxman graphs are an example of the general class of SERNs (Spatially Embedded Random Graphs) [27], and there are other related models such as the geometric inhomogeneous random graph [28], and scale-free percolation [29]. One aim of this work is to develop intuition which can be extended to estimation of the parameters of random graphs from the general class of spatial networks. We will leave consideration of the general case to later work because complex questions of existence and identifiability arise.

3   General Properties of Waxman Graphs

The starting point for the creation of a Waxman random graph is to generate a set of n points uniformly at random in some region R of a metric space Ω. For any given region we can derive a probability density function $g(t)$ for the distance between an arbitrary pair of random points. This is the Line-Picking-Problem: common instances of regions for which analytic expressions exist include lines, balls, spheres, cubes, and rectangles, *e.g.*, see [30,31].

Given the distribution of distances between points, we can calculate the probability that an arbitrary link exists (prior to knowing the distances):

$$\mathbb{P}\{(i,j) \in \mathcal{E} \mid q, s\} = q \int_0^\infty \exp(-st)g(t)\, dt = q\tilde{G}(s), \qquad (2)$$

for any $i \neq j$, where $\tilde{G}(s)$ is the Laplace transform of $g(t)$ (or equivalently it is the moment generating function w.r.t. to $-s$). We know that the Laplace transform at $s = 0$ of a probability density is the normalisation constraint, so $\tilde{G}(0) = 1$. Hence when $s = 0$ there is no distance dependence and the result is the GER random graph.

From this probability we can also compute features of the graph such as the average node degree

$$\bar{k} = (n-1)q\tilde{G}(s), \qquad (3)$$

from which we can derive values of q that produce given average degree for a given network size and s. From the Handshake Theorem we can derive the average number of edges to be

$$\bar{e} = n(n-1)q\tilde{G}(s)/2. \qquad (4)$$

We can then derive the distribution of the length d of a link in the Waxman graph, and we denote this by $f(d \mid q, s) = \mathbb{P}\{d_{ij} = d \mid (i,j) \in \mathcal{E}\}$ which is

$$f(d \mid q, s) = \frac{\mathbb{P}\{(i,j) \in \mathcal{E} \mid d_{(i,j)} = d; q, s\}\, \mathbb{P}\{d_{ij} = d\}}{\mathbb{P}\{(i,j) \in \mathcal{E} \mid q, s\}} = \frac{g(d)\exp(-sd)}{\tilde{G}(s)}. \qquad (5)$$

Note that q is a thinning parameter, and thus should not change the link length distribution, and we see that it vanishes from the distribution. Hence we generally write f omitting q, $i.e.$, $f(d \mid s)$.

We can then derive the mean length of links in the Waxman graph

$$\mathbb{E}[d \mid s] = \frac{1}{\tilde{G}(s)} \int_0^\infty tg(t) \exp(-st)\, dt = -\frac{\tilde{G}'(s)}{\tilde{G}(s)}. \tag{6}$$

We can also compute features such as the Kullback-Leibler distance between the GER graph and a given Waxman model [5].

For small distances t, region-boundary effects are minimal, and so the function $g(t)$ depends only on the dimension of the embedding space. For example the square and disk both have $g(t) \sim 2\pi t$ for small t, which comes from the size of the ring of radius t. Given a Euclidean distance metric on a k-dimensional space the small t approximation is

$$g(t) \simeq \frac{k\pi^{k/2}}{\Gamma(k/2+1)} t^{k-1}, \tag{7}$$

which is the surface area of the $(k-1)$-sphere.

Note that the distribution varies as a power-law for small t, and hence we can apply a so-called Tauberian theorem to derive the form of the tail of the Laplace transform. From [32, Theorem 2, pp. 445-6], we obtain

$$\tilde{G}(s) \overset{s\to\infty}{\sim} \frac{\pi^{k/2}\Gamma(k+1)}{\Gamma(k/2+1)} s^{-k}, \tag{8}$$

$$\tilde{G}'(s) \overset{s\to\infty}{\sim} \frac{-k\pi^{k/2}\Gamma(k+1)}{\Gamma(k/2+1)} s^{-k-1}, \tag{9}$$

$$\mathbb{E}[d \mid s] \overset{s\to\infty}{\sim} k/s, \tag{10}$$

providing some intuition regarding the average length of edges and the parameters of the graph.

The likelihood function for a particular Waxman graph under the usual independence assumption, given the lengths of the observed links $\mathbf{d}$, is

$$\mathcal{L}(s \mid \mathbf{d}) = \prod_{(i,j)\in\mathcal{E}} f(d_{i,j} \mid s). \tag{11}$$

We apply (11) below, though links in the Waxman graph are only conditionally independent given the distances. The underlying distances are a metric so three distances satisfy the triangle inequality and thus the three must be correlated. Fortunately the correlation is largely local. For instance, if we consider two distinct node pairs (i_1, j_1) and (i_2, j_2), then edges (i_1, j_1) and (i_2, j_2) are independent. Thus correlations are mediated through common nodes. As a result we should expect weaker correlations as the number of nodes n grows.

4   Estimation Techniques

4.1   Log-Linear Regression

The first method applied to this problem was introduced by [26], where is was noted that

$$\frac{f(d \mid s)}{g(d)} = c \exp(-sd), \tag{12}$$

where we can see from our calculations that $c = 1/\tilde{G}$. If we could form this ratio, we might estimate s by log-linear regression against d. The observed lengths of links in the graph yield implicit measurement of the numerator in (12) and we could obtain estimates of $g(d)$ either:

1. *analytically:* using the shape of the region to compute $g(d)$; or
2. *empirically:* using the distances between all of the nodes to estimate $g(d)$.

The former has the advantage of being fast. The latter exploits the data itself in the case that the region is not regular, but computationally it is $O(n^2)$. The latter approach was used by [26], but we shall evaluate both.

   We also need to estimate $f(d \mid s)$: [26] did so by forming a histogram. We shall use this approach, so the estimator proceeds by counting the number of edges in each length bin, and dividing by the expected number in that bin absent the distance deterrence function.

   Once we have computed $\hat{s}$, we estimate q by inverting (4) to get

$$\hat{q} = \frac{2e}{n(n-1)\tilde{G}(\hat{s})}, \tag{13}$$

where n and e are the observed number of nodes and edges respectively. The decoupling of estimation of s and q makes this sequential estimation possible, and this will be exploited in other estimators described below.

   Our evaluation of complexity focuses on the time to perform the regression, *i.e.,* we do not include the time to calculate distances, as this is not formally a component of the estimation process. Thus the time complexity using the analytical formulation of $g(d)$ is $O(e)$ but the empirical approach, which uses the distances between all nodes is $O(n^2)$.

4.2   Generalised Linear Model (GLM)

Davis *et al.* [19] use a GLM [33, p. 591] to estimate the parameters of their model. Waxman's model is slightly different, but we can use a similar approach. The response variables $Y_{(i,j)}$ are an indication of an edge so $\mathbf{Y}$ is the adjacency matrix of the graph, *i.e.,*

$$Y_{(i,j)} = \begin{cases} 1, & \text{if } (i,j) \in \mathcal{E}, \\ 0, & \text{otherwise.} \end{cases} \tag{14}$$

The predictor variables $\mathbf{X}$ for links are the distances, and the model is

$$r(\mu_{(i,j)}) = \beta_0 + \beta_1 X_{(i,j)}, \tag{15}$$

where $\mu_{(i,j)} = \mathbb{E}[Y_{(i,j)}] = p(d_{i,j})$, and $r(\cdot)$ is the link function. The GLM of [19] used the link function $r(p_{(i,j)}) = \log(-\log(1 - p_{(i,j)}))$. Here, because of the slightly simpler model, we use $r(\mu) = \log(\mu)$, which facilitates the natural identification that $\beta_0 = \log(q)$ and $\beta_1 = -s$. Estimation of the β_m is usually by maximum likelihood (note however, that we distinguish the GLM from our MLE defined below), and we use Matlab's `glmfit` to perform this task.

The GLM method uses all possible edges: those that exist and those that do not. *i.e.*, there are $n(n-1)/2$ response variables, and so the computational complexity is $\Omega(n^2)$.

4.3 Sufficient Statistics

An obvious question arises as to what is a set of *minimal sufficient statistics* for use in the estimation of Waxman random graph parameters. The following theorem answers this.

Theorem 1. *The number of edges e and their average length $\bar{d}$ form minimal sufficient statistics for the parameters of a Waxman random graph.*

Proof. Theorem 6.2.13 of [33] states that given the PDF $f(\mathbf{x} \mid \theta)$ of a sample $\mathbf{X}$ and a function $T(\mathbf{x})$ such that, for every two sample points $\mathbf{x}$ and $\mathbf{y}$, the ratio $f(\mathbf{x} \mid \theta)/f(\mathbf{y} \mid \theta)$ is a constant function of θ if and only if $T(\mathbf{x}) = T(\mathbf{y})$. Then $T(\mathbf{X})$ is a minimal sufficient statistic for θ.

Take the sample to be the set of edge lengths $\mathbf{d} = (d_1, \ldots, d_e)$ where $d_k = d_{(i,j)}$ for $(i,j) \in \mathcal{E}$. Under the independence assumption the PDF of a sample is $\prod_{i=1}^{e} f(d_i|s,q)$, conditional on the number of edges e, where f is defined in (5), and e is binomially distributed $B_p^N(e)$ where $N = n(n-1)/2$ and $p = q\tilde{G}(s)$. Therefore for two samples: $\mathbf{x}$ and $\mathbf{y}$, the ratio of PDFs is

$$\frac{B_p^N(e_1) \prod_{i=1}^{e_1} f(x_i|s,q)}{B_p^N(e_2) \prod_{i=1}^{e_2} f(y_i|s,q)} = m(\mathbf{x},\mathbf{y}) \exp\left(-s(e_1\bar{d}_x - e_2\bar{d}_y)\right) \tilde{G}(s)^{e_2-e_1} \frac{B_p^N(e_1)}{B_p^N(e_2)},$$

where $\bar{d}_x$ and $\bar{d}_y$ are the averages of the distances in datasets $\mathbf{x}$ and $\mathbf{y}$ respectively, and $m(\mathbf{x},\mathbf{y})$ is independent of the parameters (q,s).

The ratio above depends on the parameters s and q only through the statistics $\bar{d}$ and e. Hence if $e_1 = e_2$ and $\bar{d}_x = \bar{d}_y$, the ratio is a constant function of (q,s). On the other hand, if either $e_1 \neq e_2$ or $\bar{d}_x \neq \bar{d}_y$, then the ratio is a non-constant function of the parameters. Hence the conditions of Theorem 6.2.13 are satisfied and $(e, \bar{d})$ forms a minimal sufficient set of statistics. $\qquad\square$

Notably, these statistics can be constructed almost trivially in $O(e)$ time, and we will use that fact to construct a MLE.

4.4   Maximum Likelihood Estimator

Using (11), the log-likelihood function $\ell(s \mid \mathbf{d}) = \ln \mathcal{L}(s \mid \mathbf{d})$ can be shown to be

$$
\begin{aligned}
\ell(s \mid \mathbf{d}) &= \sum_{(i,j)\in\mathcal{E}} \ln f(d_{(i,j)} \mid s) \\
&= \sum_{(i,j)\in\mathcal{E}} \left[\ln g(d_{(i,j)}) - sd_{(i,j)} - \ln \tilde{G}(s) \right] \\
&= -e \ln \tilde{G}(s) + \sum_{(i,j)\in\mathcal{E}} \ln g(d_{(i,j)}) - s \sum_{(i,j)\in\mathcal{E}} d_{(i,j)}.
\end{aligned}
$$

Our goal is to find s such that the partial derivative of $\ell(s \mid \mathbf{d})$ with respect to s is zero, *i.e.*,

$$
\frac{\partial}{\partial s}\ell(s \mid \mathbf{d}) = -e\frac{\tilde{G}'(s)}{\tilde{G}(s)} - \sum_{(i,j)\in\mathcal{E}} d_{(i,j)} = 0, \tag{16}
$$

so we need to find s such that

$$
\frac{\tilde{G}'(s)}{\tilde{G}(s)} = -\frac{1}{e} \sum_{(i,j)\in\mathcal{E}} d_{(i,j)} = -\bar{d}, \tag{17}
$$

where $\bar{d}$ is the observed mean link length.

   From (6) we know $-\tilde{G}'/\tilde{G}$ is the expected length of line segments on the Waxman graph, so the MLE of s is also the moment-matching estimator.

4.5   Existence and Uniqueness of the MLE

When will a unique solution to (17) exist? We know from the definitions of $\tilde{G}(s)$ and its derivative that $\tilde{G}(s) > 0$, and $h(s) = -\tilde{G}'(s)/\tilde{G}(s)$ is a continuous, positive function for all s, so if $h(\cdot)$ is monotonic there will be at most one solution to (17) for any given $\bar{d}$.

   Consider the derivative

$$
\frac{dh}{ds} = \frac{\tilde{G}'(s)^2 - \tilde{G}''(s)\tilde{G}(s)}{\tilde{G}(s)^2}. \tag{18}
$$

Now the numerator is positive, and from Schwarz's Inequality

$$
\begin{aligned}
\tilde{G}''(s)\tilde{G}(s) &= \int \left| \sqrt{t^2 g(t)e^{-st}} \right|^2 dt \cdot \int \left| \sqrt{g(t)e^{-st}} \right|^2 dt \\
&> \left| \int tg(t)e^{-st} \, dt \right|^2 = \tilde{G}'(s)^2,
\end{aligned}
$$

and so the denominator is negative, and the function $h(s)$ is monotonically decreasing with s.

When $s = 0$, the Laplace transform $\tilde{G}(0) = 1$ and hence $h(0) = \int t g(t) dt = \bar{g}$, the average distance between pairs of nodes. When we remove longer links preferentially, the average link distance must decrease, so it is intuitive that $h(s)$ is a decreasing function, starting at $h(0) = \bar{g} = d_{max}$, which is the maximum expected edge distance over all possible parameters s.

In the limit as $s \to \infty$ from (10) we have $h(s) = \mathbb{E}[d] \simeq k/s$, and so we know that in the limit as $s \to \infty$ that $h(s) \to 0$, so for any measured $\bar{d} \in (0, d_{max}]$ there will be a unique solution to (17).

Unfortunately, it is possible for the sample mean of the edge distances $\bar{d} > d_{max}$, i.e., for a particular graph to have an unlikely preponderance of longer links. In this case (17) has no solution for $s \in [0, \infty)$. However, the obvious interpretation of $\bar{d} > d_{max}$ is that there is no evidence that long links have been preferentially filtered from the graph, and so it is natural in this case to assume the model should be the GER random graph, i.e., $s = 0$.

In formal terms, the MLE satisfies some but not all of the properties required for it to be *consistent*. Standard asymptotic theory for MLEs requires the condition that the true parameter value lies away from the boundary to form consistent estimates that converge to the true value as the amount of data increases. Surprisingly, for cases near $s = 0$ this can induce a small bias leading to a variance-bias tradeoff and an RMS error in the estimator that can be lower than the Cramér-Rao (CR) bound. Apart from this case the MLE is *asymptotically normal*, i.e., as the sample size increases, the distribution of the MLE tends to the Gaussian distribution with mean given by the true parameter, and covariance matrix equal to the inverse of the Fisher information matrix. For more discussion and derivation of the CR bound see [5].

4.6   The MLE of $\hat{q}$

As noted above, once we know $\hat{s}$, we can use this to estimate $\hat{q}$ using (13). If we know the exact value of s then q acts as a random filter of links, so estimation of q is equivalent to estimating the parameter of a Bernoulli process. Thus (13) would be the MLE of q given the true value of s. The question is whether it is still a MLE when we derive it from the estimated value of $\hat{s}$.

To answer this question we draw on the property of functional invariance of MLE, i.e., if parameters are related through a transformation $a = g(s)$, then the MLE of a is $\hat{a} = g(\hat{s})$. This property allows the conversion back to the original Waxman parameter $\alpha = 1/sL$, for instance. More importantly it means that $\hat{q} = 2e/n(n-1)\tilde{G}(\hat{s})$ is the MLE of q, given a MLE of $\hat{s}$.

4.7   Numerical Calculation of the MLE

MLEs are often computed directly using techniques such as Newton's method. In this problem, each iteration would require numerical computation of a Laplace transform and its derivative. However, we can improve this in several ways. First, we rearrange (17) in a form that halves the required number of transforms

$$\bar{d}\tilde{G}(s) + \tilde{G}'(s) = \mathcal{LT}\left[(\bar{d} - t)g(t); s\right] = 0. \tag{19}$$

Second, the estimation algorithm is a 1D search. Its speed can be improved by providing reasonable bounds: here we use $[0, k/\bar{d}]$, where the upper bound is given by the asymptotic form of the average distance (10).

Third, we can calculate transforms using adaptive numerical quadrature, which is the most accurate approach. However, if we precalculate $g(t)$ at a set of fixed grid points, and reuse this in all of the Laplace transform calculations, it is faster. We call this the MLE-N (Numerical).

The grid used in the MLE-N samples the distribution at uniformly spaced probabilities, which also means that we can use an estimate of the inverse CDF derived from data based on the complete set of distances between all nodes and derive from this an empirical estimate of the inverse CDF to use in the MLE calculations. We refer to this method as the MLE-E (Empirical).

5   Performance

In this section we examine the performance of the Waxman graph parameter estimators on simulated data. We primarily focus on fairly sparse networks ($\bar{k} = 3$) because sparsity is the common case for many real networks, but we do examine the effect of changes in these parameters.

The accuracy of the methods increase as the graphs become larger so we start with parameter values that result in moderately sized (1000 node) graphs. This ensures that errors are of a magnitude that allows us to compare and assess them. However, we also examine the relationship between graph size and accuracy of parameter estimates.

We simulate 1000 Waxman random graphs for each parameter setting for the Waxman graph on the unit square with the Euclidean distance metric. Other cases and estimator robustness are considered in more detail in [5].

5.1   Comparisons

We start by considering the RMS errors of the methods with respect to the CR lower bound as a function of the parameter s. Figure 1 shows these errors. The most obvious difference is between the log-linear and other approaches, the log-linear estimator being much less accurate. Focusing on the group of "good" approaches, the entire range the GLM tracks the CR bound, but for small values of s there is some advantage in using the GLM and MLE-E. For very small s, the MLE approaches can actually beat the CR bound. This can be explained by the constraint $s \geq 0$. This constraint leads to a small bias in the MLE estimators for small s. When s is around 1, the bias leads to an increase in the RMS errors. When s is very small the estimators have slightly more information than assumed in our naive CR bound, and this can reduce the variance of the estimator [34].

For large values of s the MLE-E and MLE-N show a small increase in error compared to the exact MLE, because for large s the grid size chosen is not small enough for accurate integrals to be calculated. This effect could be mitigated by choosing a finer grid or by choosing a non-uniform grid with more detail around $t = 0$ (albeit at additional computational cost), but it is quite a small effect.

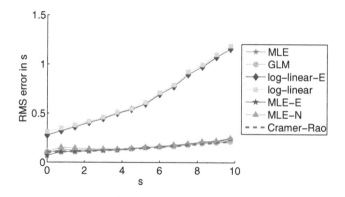

Fig. 1. RMS errors as a function of s ($\bar{k} = 3$, $n = 1000$).

For small values of s, the MLE-E estimator is slightly better than its exact cousin. This is valuable, but remember that the MLE-E requires the information on all distances, not just those of the existing edges, whereas the MLE and MLE-N estimators need only the measurement of $\bar{d}$. For large graphs $\bar{d}$ could be calculated from a sample of edges, so it is possible to achieve sub-$O(e)$ performance for such graphs.

We have examined a number of parameter settings and have observed the same trends in the results – for other instances and more detailed plots see [5], which also shows that the majority of error is in the form of bias.

In summary: the log-linear regression is the least accurate and that the GLM and MLE estimators are all roughly equivalent in accuracy. The minor empirical variants of the log-linear and MLE estimators have only small effects, typically at extreme values.

We also estimate computation times using Matlab's *tic/toc* timers to estimate wall-clock time of execution, and take the median over our 1000 samples to provide a robust estimate of the typical computation times, which are shown in Fig. 2. We can see that the GLM takes quadratic time, asymptotically. The log-linear-E method (not shown) is also quadratic, because a histogram must be formed from all of the distances, but it is significantly faster than the GLM.

The other methods appear to have constant time with respect to the network size n. Constant time is an illusion however: these methods are actually $O(e)$, which for the networks considered is also $O(n)$. The constant time component is obviously dominant for the network sizes considered – we would expect to see the linearity only if we examine very large networks.

Memory requirements for the algorithms are

- GLM: $O(n^2)$;
- log-linear (and -E variant): $O(b)$ where b is the number of histogram bins;
- MLE: $O(1)$ (assuming the numerical quadrature is memory efficient); and
- MLE-E and MLE-N: $O(c)$ where c is the number of grid points.

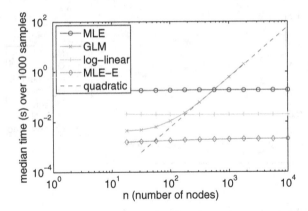

Fig. 2. Computation times ($s = 4$, $\bar{k} = 3$). Note that only the estimator time is included, not the time to form the matrix of distances needed in the GLM and log-linear methods.

All but the GLM use constant memory as a function of network size, as they are based on summary statistics that can be computed via a streaming algorithm.

If accuracy is the prime consideration then the GLM is the best approach, however, for large networks it is impractical. Extrapolating the computation times for the GLM shown in Fig. 2, we can see that computation would take in the order of 3 h for a graph with 100,000 nodes. And the GLM requires a large amount of memory for larger networks, as compared with the almost trivial amount required in the other algorithms. The log-linear approaches are poor on all fronts. Variants of the MLE approach are faster, and more accurate.

We now consider the accuracy of the MLE as both a function of the network size n in Fig. 3(a) and of the average node degree $\bar{k}$ in Fig. 3(b). The accuracy of the method is close to the CR bound, but suffers slightly for small networks, and for higher average node degree. This variation is explained by deviations in the independence assumption.

5.2   Estimating q

In most of the preceding work we have only considered the accuracy of the estimates for s. The estimated accuracy of q is that of the binomial model parameter estimate which is well understood. There is an additional source of error in that errors in $\hat{s}$ propagate into estimate $\hat{q}$ through $\tilde{G}(\hat{s})$. This roughly doubles the size of errors in q compared to estimates based on the correct value of s, but the errors in $\hat{q}$ are still small: *i.e.*, of the order of 5% (see [5] for details).

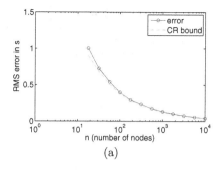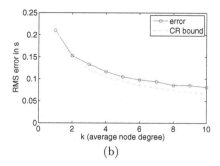

(a)   (b)

Fig. 3. Accuracy of the MLE as a function of other parameters of the network (default parameters $n = 1000$, $s = 3.0$, and $\bar{k} = 3$).

6   Case Study

The previous section looked at accuracy on simulated data, where we know the ground truth. This section applies the method to a real dataset from the Internet examining the length of physical links between routers.

Lakhina *et al.* [26] undertook one of the first attempts to formally quantify the exponential decrease of link likelihood as a function of distance. The authors compared two sets of data and found consistent results between them. They provided one of these datasets (the Mercator data) to us for comparison. Lakhina *et al.* separated the data into three regions (the USA, Europe and Japan), and analysed these separately, which we repeat here. These graphs are large, with up to 123,000 nodes, but very sparse, with average node degrees of around 2. It is not feasible to use the GLM approach for this scale of problem.

We do not argue that this network is random in any real sense: in fact the Internet networks are the result of design. However, fitting a Waxman-like graph to these is instructive in that it shows how engineering constraints lead to distance-sensitive link placement.

Lakhina *et al.* [26] applied log-linear regression to the question. We applied the MLE and compared the results to those found in their work. Table 1 provides a comparison between various estimates, including the original values reported in [26] given in the second column under $\hat{s}_{Lakhina}$. Units are *per 1000 miles*—we use Imperial units to be consistent with the original paper. The third column provides our equivalent estimate. There are small differences, presumably because of differences in the exact numerical procedures applied. The fourth column of the table shows the MLE-E values estimated for the datasets. We use the Empirical estimator because the region shapes are irregular (*e.g.*, the USA), and we want to avoid approximation errors arising from the region shape.

We see considerable discrepancies which are larger than can be explained by errors in the log-linear regression approach. However, reading [26], we can see that their estimates are over a truncated range of distances for two reasons:

Table 1. Estimates: $\hat{s}_{Lakhina}$ are the values from [26]; and $\hat{s}_{log-linear}$ are our corresponding estimates. Units are *per 1000 miles*. The $\hat{s}_{MLE-E}$ values are derived from our Empirical MLE, and the $\hat{s}_{MLE-T}$ values from a version of the MLE with distance data truncated in the same manner as the original log-linear estimates of [26]

Region	$\hat{s}_{Lakhina}$	$\hat{s}_{log-linear}$	$\hat{s}_{MLE-E}$	$\hat{s}_{MLE-T}$
USA	6.91	6.38	2.75	6.63
EUROPE	12.80	12.81	30.92	10.09
JAPAN	6.89	6.71	45.91	7.30

- The node locations they use are artificially quantised by the Geolocation procedure so some nodes appear to have exactly the same position, and hence zero distance, when actually there is a positive distance between the nodes.
- They found that the exponential distance-deterrence function fit the data only up to some threshold distance.

In their (and our comparison) log-linear regressions the range over which we perform the regression is restricted to be between these bounds.

In order to provide a fair comparison we also modified the MLE-E by censoring the potential edges used in forming the CDF and in computing the average edges distance. Table 1 shows the results under s_{MLE-T} to be closer to being consistent with the log-linear regression.

The results point to one valuable feature of the log-linear regression, which is that it comes with diagnostics. Examination of the fit indicates whether the model is appropriate or not. The MLE requires additional effort to provide similar diagnostics. On the other hand, there are significant issues with the log-linear regression. Apart from being less accurate, there is the question of bin size which simple experiments seem to suggest is important.

Ultimately, all of the methods suggest strongly that a spatial component should be part of any model for Internet linkages. This is entirely consistent with the intuition of engineers who work on such networks: long links cost more, and so are rarer.

7   Discussion and Conclusion

This paper presents the MLE for the parameters of the Waxman graph and demonstrates its accuracy in comparison to alternative estimators. The MLE has two advantages. Firstly it has $O(e)$ computational time complexity and constant memory usage by using only a sample of the edges that exist in a graph to estimate s. Secondly it can be applied in domains where the coordinates of nodes are unknown and/or edge lengths may be weights in some arbitrary process.

Acknowledgements. We would like to thank Lakhina *et al.* for providing the Internet dataset.

References

1. Waxman, B.: Routing of multipoint connections. IEEE J. Select. Areas Commun. **6**(9), 1617–1622 (1988)
2. Gilbert, E.: Random graphs. Ann. Math. Stat. **30**, 1141–1144 (1959)
3. Erdős, P., Rényi, A.: On the evolution of random graphs. Publ. Math. Inst. Hung. Acad. Sci. **5**, 17–61 (1960)
4. Fosdick, B.K., Hoff, P.D.: Testing and modeling dependencies between a network and nodal attributes. ArXiv e-prints (2013). http://arxiv.org/abs/1306.4708
5. Roughan, M., Tuke, J., Parsonage, E.: Estimating the parameters of the Waxman random graph. ArXiv e-prints (2015). http://arxiv.org/abs/1506.07974
6. Zegura, E.W., Calvert, K., Bhattacharjee, S.: How to model an internetwork. In: IEEE INFOCOM, CA, San Francisco, pp. 594–602 (1996)
7. Salama, H.F., Reeves, D.S., Viniotis, Y.: Evaluation of multicast routing algorithms for real-time communication on high-speed networks. IEEE J. Sel. Areas Commun. **15**, 332–345 (1997)
8. Verma, S., Pankaj, R.K., Leon-Garcia, A.: QoS based multicast routing algorithms for real time applications. Perform. Eval. **34**, 273–294 (1998)
9. Shaikh, A., Rexford, J., Shin, K.G.: Load-sensitive routing of long-lived IP flows. In: ACM SIGCOMM (1999)
10. Neve, H.D., Mieghem, P.V.: TAMCRA: a tunable accuracy multiple constraints routing algorithm. Comput. Netw. **23**, 667–679 (2000)
11. Wu, J.-J., Hwang, R.-H., Lu, H.-I.: Multicast routing with multiple QoS constraints in ATM networks. Inf. Sci. **124**(1–4), 29–57 (2000). http://www.sciencedirect.com/science/article/pii/S0020025599001024
12. Guo, L., Matta, I.: Search space reduction in QoS routing. Comput. Netw. **41**, 73–88 (2003)
13. Gunduz, C., Yener, B., Gultekin, S.H.: The cell graphs of cancer. Bioinformatics **20**(1), 145–151 (2004)
14. Carzaniga, A., Rutherford, M.J., Wolf, A.L.: A routing scheme for content-based networking. In: IEEE INFOCOM (2004)
15. Holzer, M., Schulz, F., Wagner, D., Willhalm, T.: Combining speed-up techniques for shortest-path computations. J. Exp. Algorithmics **10**, 2–5 (2005)
16. Malladi, S., Prasad, S., Navathe, S.: Improving secure communication policy agreements by building coalitions. In: IEEE Parallel and Distributed Processing Symposium, pp. 1–8, March 2007
17. Tran, D.A., Pham, C.: PUB-2-SUB: a content-based publish/subscribe framework for cooperative P2P networks. In: Fratta, L., Schulzrinne, H., Takahashi, Y., Spaniol, O. (eds.) NETWORKING 2009. LNCS, vol. 5550, pp. 770–781. Springer, Heidelberg (2009). https://doi.org/10.1007/978-3-642-01399-7_60
18. Fang, Y., Chu, F., Mammar, S., Che, A.: Iterative algorithm for lane reservation problem on transportation network. In: 2011 IEEE International Conference on Networking, Sensing and Control (ICNSC), pp. 305–310, April 2011
19. Davis, S., Abbasi, B., Shah, S., Telfer, S., Begon, M.: Spatial analyses of wildlife contact networks. J. R. Soc. Interface **12**(102) (2014). http://rsif.royalsocietypublishing.org/content/12/102/20141004.short
20. Zegura, E.W., Calvert, K.L., Donahoo, M.J.: A quantitative comparison of graph-based models for Internet topology. IEEE/ACM Trans. Netw. **5**(6), 770–783 (1997)
21. Van Mieghem, P.: Paths in the simple random graph and the Waxman graph. Probab. Eng. Inf. Sci. **15**, 535–555 (2001). http://dl.acm.org/citation.cfm?id=982639.982646

22. Naldi, M.: Connectivity of Waxman graphs. Comput. Commun. **29**, 24–31 (2005)
23. Snijders, T.A.B., Koskinen, J., Schweinberger, M.: Maximum likelihood estimation for social network dynamics. Ann. Appl. Stat. **4**(2), 567–588 (2010). https://doi. org/10.1214/09-AOAS313
24. Frank, O., Strauss, D.: Markov graphs. J. Am. Stat. Assoc. **81**(395), 832–842 (1986)
25. Robins, G., Pattison, P., Kalish, Y., Lusher, D.: An introduction to exponential random graph (p*) models for social networks. Soc. Netw. **29**(2), 173–191 (2007). http://www.sciencedirect.com/science/article/pii/S0378873306000372
26. Lakhina, A., Byers, J.W., Crovella, M., Matta, I.: On the geographic location of Internet resources. In: ACM SIGCOMM Workshop on Internet Measurement (IMW), pp. 249–250. ACM, New York (2002). https://doi.org/10.1145/637201. 637240
27. Kosmidis, K., Havlin, S., Bunde, A.: Structural properties of spatially embedded networks. Europhys. Lett. **82**(4) (2008). http://iopscience.iop.org/0295-5075/82/ 4/48005
28. Bringmann, K., Keusch, R., Lengler, J.: Geometric inhomogeneous random graphs. ArXiv e-prints, November 2015. https://arxiv.org/abs/1511.00576
29. Deijfen, M., van der Hofstad, R., Hooghiemstra, G.: Scale-free percolation. ArXiv e-prints, March 2011. https://arxiv.org/abs/1103.0208
30. Ghosh, B.: Random distance within a rectangle and between two rectangles. Bull. Calcutta Math. Soc. **43**(1), 17–24 (1951)
31. Rosenberg, E.: The expected length of a random line segment in a rectangle. Oper. Res. Lett. **32**(2), 99–102 (2004). http://www.sciencedirect.com/science/ article/pii/S0167637703000725
32. Feller, W.: An Introduction to Probability Theory and its Applications, vol. II, 2nd edn. Wiley, New York (1971)
33. Casella, G., Berger, R.L.: Statistical Inference, 2nd edn. Thomson Learning, Pacific Grove (2002)
34. Gorman, J.D., Hero, A.O.: Lower bounds for parametric estimation with constraints. IEEE Trans. Inf. Theory **36**(6), 1285–1301 (1990)

Understanding the Effectiveness of Data Reduction in Public Transportation Networks

Thomas Bläsius, Philipp Fischbeck[(✉)], Tobias Friedrich, and Martin Schirneck

Hasso Plattner Institute, University of Potsdam, Potsdam, Germany
{thomas.blaesius,philipp.fischbeck,tobias.friedrich,
martin.schirneck}@hpi.de

Abstract. Given a public transportation network of stations and connections, we want to find a minimum subset of stations such that each connection runs through a selected station. Although this problem is NP-hard in general, real-world instances are regularly solved almost completely by a set of simple reduction rules. To explain this behavior, we view transportation networks as hitting set instances and identify two characteristic properties, locality and heterogeneity. We then devise a randomized model to generate hitting set instances with adjustable properties. While the heterogeneity does influence the effectiveness of the reduction rules, the generated instances show that locality is the significant factor. Beyond that, we prove that the effectiveness of the reduction rules is independent of the underlying graph structure. Finally, we show that high locality is also prevalent in instances from other domains, facilitating a fast computation of minimum hitting sets.

Keywords: Transportation networks · Hitting set ·
Graph algorithms · Random graph models

1 Introduction

A public transportation network is a collection of stations along with a set of connections running through these stations. But beyond its literal definition, via bus stops and train lines, it also carries some of the geographical, social, and economical structure of the community it serves. Given such a network, we want to select as few stations as possible to *cover* all connections, i.e., each connection shall contain a selected station. This and similar covering problems arise from practical needs, e.g., when choosing stations for car maintenance, but their solutions also reveal some of the underlying structure of the network. Despite the fact that minimizing the number of selected stations is NP-hard, there is a surprisingly easy way to achieve just that on real-world instances: Weihe [20] showed for the German railroad network that two straightforward reduction rules simplify the network to a very small core which can then be solved by brute force. This

© Springer Nature Switzerland AG 2019
K. Avrachenkov et al. (Eds.): WAW 2019, LNCS 11631, pp. 87–101, 2019.
https://doi.org/10.1007/978-3-030-25070-6_7

is not a mere coincidence. Experiments have shown the same behavior on several other real-world transportation networks. Subsequently, the reduction rules became the standard preprocessing routine for many different covering problems. See the work of Niedermeier and Rossmanith [13], Abu-Khzam [1], or Davies and Bacchus [6], to name just a few. This raises the question as to why these rules are so effective. Answering this question would not only close the gap between theory and practice for the specific problem at hand, but also has the potential to lead to new insights into the networks' structure and ultimately pave the way for algorithmic advances in bordering areas.

Our methodology for approaching this question is as follows. We first identify two characteristic properties of real-world transportation networks: *heterogeneity* and *locality*; see Sect. 2.2 for more details. Then we propose a model that generates random instances resembling real-world instances with respect to heterogeneity and locality. We validate our model by showing empirically that it provides a good predictor for the effectiveness of the reduction rules on real-world instances. Finally, we draw conclusions on why the reduction rules are so effective by running experiments on generated instances of varying heterogeneity and locality. Moreover, we show that our results extend beyond transportation networks to related problems in other domains.

For our model, we regard transportation networks as instances of the *hitting set problem*. From this perspective, connections are mere subsets of the universe of stations and we need to select one station from each set. Note that this disregards some of the structure inherent to transportation networks: A connection is not just a set of stops but a sequence visiting the stops in a particular order. In fact, the sequences formed by the connections are paths in an underlying graph, which itself has rich structural properties inherited from the geography. Focusing on these structural properties, we also consider the *graph-theoretic perspective*. The working hypothesis for this perspective is that the underlying graphs of real-world transportation networks have beneficial properties that render the instances tractable. We disprove this hypothesis by showing that the underlying graph is almost irrelevant. This validates the hitting set perspective, which disregards the underlying graph.

In Sect. 2, we formally state our findings on the graph-theoretic as well as the hitting set perspective. We study the hitting set instances of European transportation networks in Sect. 3, identifying heterogeneity and locality as characteristic features. In Sect. 4, we define and evaluate a model generating instances with these features. Section 5 extends our findings to other domains and Sect. 6 concludes this work.

2   Preliminary Considerations

Before discussing the results regarding the two different perspectives, we fix some notations and state the reduction rules introduced by Weihe [20]. A *public transportation network* (or simply a *network*) $N = (S, C)$ consists of a set S of *stations* and a set C of *connections* which are sequences of stations. That is, each connection $c \in C$ is a subset of S together with a linear ordering of its elements. Two stations $s_1, s_2 \in S$ are *connected* in N if there exists a sequence of stations starting with s_1 and ending in s_2 such that each pair of consecutive stations shares a connection. The subnetworks induced by this equivalence relation are called the *connected components* of N. Given $N = (S, C)$, the STATION COVER problem is to find a subset $S' \subseteq S$ of minimum cardinality such that each connection is *covered*, i.e., $S' \cap c \neq \emptyset$ for every $c \in C$. The reduction rules by Weihe [20] are based on notions of dominance, both between stations and connections. For two different stations $s_1, s_2 \in S$, s_1 *dominates* s_2 if every connection containing s_2 also contains s_1. If so, there is always an optimal station cover without s_2, so it is never worse to select s_1 instead. Thus, removing s_2 from S and from every connection in C yields an equivalent instance. Similarly, for two different connections $c_1, c_2 \in C$, c_1 *dominates* c_2 if $c_1 \subseteq c_2$. Every subset of S covering c_1 then also covers c_2. Removing c_2 does not destroy any optimal solutions. Weihe's algorithm can thus be summarized as follows. Iteratively remove dominated stations and connections until this is no longer possible. The remaining instance, the *core*[1], is solved using brute force. Each connected component can be solved independently and the running time is exponential only in the number of stations. Thus, the *complexity* of an instance denotes the maximum number of stations in any of its connected components.

The proofs of this section can be found in the full version of this paper [4].

2.1   Graph-Theoretic Perspective

One way to represent a network $N = (S, C)$ is via an undirected graph G_N defined as follows. The stations S are the vertices of G_N; for each connection $(s_1, \ldots, s_k) \in C$, G_N contains the edges $\{s_i, s_{i+1}\}_{1 \leq i < k}$. The basic hypothesis of the graph-theoretic perspective is that certain properties of G_N make the real-world STATION COVER instances easy.

Consider a leaf u in G_N, i.e., a degree-1 vertex. If there is a connection that contains only u, then this dominates all other connections containing u. Otherwise, all connections that contain u also contain its unique neighbor. Thus, u is dominated and removed by the reduction rules. We obtain the following proposition. The *2-core* is the subgraph obtained by iteratively removing leaves [16].

Proposition 1. *The reduction rules reduce any* STATION COVER *instance N to an equivalent instance N' such that $G_{N'}$ is a subgraph of the 2-core of G_N, with additional isolated vertices.*

Proposition 1 identifies the number of vertices in the 2-core of G_N as an upper bound for the *core complexity*. The following theorem shows that this bound is arbitrarily bad. Supporting this assessment, we will see in Sect. 3 that the 2-cores of the graphs of real-world instances are rather large, while their core complexity is significantly smaller.

---

[1] We note that the core is unique up to automorphisms. In particular, its size is independent of the removal order.

Theorem 1. *For every graph G, there exist two* STATION COVER *instances N_1 and N_2 with $G = G_{N_1} = G_{N_2}$ such that the core of N_1 has complexity 1 while the core of N_2 corresponds to the 2-core of G.*

Theorem 1 disproves the working hypothesis of the graph-theoretic perspective. For any connected graph that has no leaves, there is a STATION COVER instance that is completely solved by the reduction rules, and another instance on the very same graph that is not reduced at all. Furthermore, unless the 2-core is small, the theorem shows that it is impossible to tell whether or not the reduction rules are effective on a given instance by only looking at the graph.

So far, we have only focused on Weihe's algorithm. While our main goal is to explain the performance of this algorithm, one could argue that other methods exploiting different graph-theoretic properties are better suited to solve real-world instances. The next theorem, however, indicates that this is not the case. Even on "tree-like" graph classes STATION COVER remains NP-hard. The reduction used to prove this theorem was originally given by Jansen [9].

Theorem 2 ([9], Theorem 5). STATION COVER *is NP-hard even if the corresponding graph has treewidth 3 or feedback vertex number 2.*

2.2   Hitting Set Perspective

Another way to represent a network N is by an instance of the HITTING SET problem. Here, the connections $C \subseteq 2^S$ are regarded only as sets of stations (ignoring their order). An optimal cover is a minimum-cardinality subset of S that has a non-empty intersection with all members of C. This perspective turns out to be much more fruitful. In the next section, we analyze the HITTING SET instances stemming from 12 real-world networks. To summarize our results, we observe that the instances are heterogeneous, i.e., the number of connections containing a given station varies heavily. Moreover, the instances exhibit a certain locality, which probably has its origin in the stations' geographic positions.

In more detail, for a station $s \in S$, let the number of connections in C that contain s be the *degree* of s. Conversely, for $c \in C$, $|c|$ is its *degree*. The connection degrees of the real-world instances are rather homogeneous, i.e., every connection has roughly the same size. Although there are different types of connections they appear to have a similar number of stops. The station degrees, on the other hand, vary strongly. In fact, we observe that the station degree distributions roughly follow a power law. This is in line with observations that, e.g., the sizes of cities are power-law distributed [7]. To quantify the locality of an instance, we use a variant of the so-called *bipartite clustering coefficient* [15].

We conjecture that heterogeneity of stations and locality of the network are the crucial factors that make the reduction so effective. If the station degrees vary strongly, chances are that some high-degree station exists that dominates many low-degree ones. Moreover, if locality is high, there tend to be several connections differing only in few stations and stations appearing in similar sets of connections. This increases the likelihood of dominance among the elements

Table 1. Transportation networks with atypical instances separated. Shown are the number $|S|$ of stations, the station-connection ratio $|S|/|C|$, average station degree δ_S, estimated power-law exponent β, corresponding KS distance, bipartite clustering coefficient κ, relative 2-core size, and the relative core complexity.

Data set	$|S|$	$\frac{|S|}{|C|}$	δ_S	β	KS	κ	2-core	Core
sncf	1742	4.0	2.2	3.3	0.03	0.47	70%	0.3%
nl	4558	13.2	1.5	3.8	0.04	0.40	70%	2.8%
kvv	2115	8.0	2.1	3.5	0.03	0.48	72%	0.8%
vrs	5491	10.7	1.9	3.5	0.03	0.27	83%	0.1%
rnv	705	12.4	1.4	4.2	0.06	0.38	54%	0.1%
athens	5729	24.4	1.8	3.9	0.04	0.30	89%	4.7%
petersburg	4264	6.5	2.5	4.0	0.03	0.31	86%	8.3%
warsaw	3944	13.0	1.8	5.9	0.05	0.29	80%	5.9%
luxembourg	2484	7.3	2.7	2.9	0.02	0.25	84%	0.2%
switzerland	22535	5.6	2.0	4.5	0.02	0.33	71%	1.7%
vbb	3031	16.5	1.4	12.4	0.05	0.38	73%	1.8%
db	514	0.9	15.7	2.0	0.07	0.28	78%	0.2%

of both S and C. To verify this hypothesis empirically, we propose a model for generating instances of varying heterogeneity and locality. Our findings suggest that higher heterogeneity decreases the core complexity, but the deciding factor is the locality. Finally, we observe that locality is also prevalent in other domains. As predicted by our model, preprocessing also greatly reduces these instances.

3 Analysis of Real-World Networks

We examined several public transportation networks from different cities (athens, petersburg, warsaw), rural areas (sncf, kvv, vrs, rnv, vbb), and countries (nl, luxembourg, switzerland, db). The networks are taken from the transitfeeds.com repository. The raw data has the General Transit Feed Specification (GTFS) format. It stores multiple connections for the same route, one for each time a train actually drives that route. For each route, only one connection was used. Table 1 gives an overview of the relevant features of the resulting networks.

We reduced each instance to its largest component. For most of them, only a small fraction of stations and connections are disconnected from this component. A notable exception is the vbb-instance, representing the public transportation network of the city of Berlin, Germany. In total, it has 13 424 stations while its largest component has only 3031. The reason is that different modes of transport are separated in the raw data. As a result, vbb has rather uncommon features. Another unusual case is the db-instance of the German railway network. Table 1

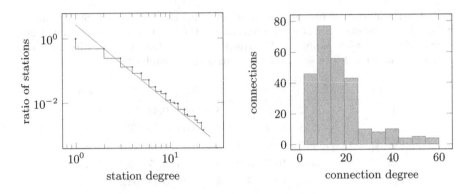

Fig. 1. (left) The blue line is the CCDF of station degrees for the data set kvv. The red line is the estimated power-law distribution. (right) The histogram of connection degrees for the data set kvv. (Color figure online)

shows that most instances have a station-connection ratio $|S|/|C|$ of roughly 10. For db, however, this ratio is at 0.9 much smaller.

Heterogeneity. The average station degree δ_S of the investigated networks is a small constant around 2, independent of the instances' complexity. The only exception is the db-network. This can be explained by the atypical value for $|S|/|C|$, and that each station is contained in much more connections. The average connection degree $\delta_C = \delta_S \cdot |S|/|C|$ (not explicitly given in the table) is roughly 20, due to the station-connection ratios all being of the same order.

Beyond small average degree, almost all instances exhibit strong heterogeneity among the degrees of different stations. We take a closer look at the kvv-instance as a prototypical example, representing the public transportation network of Karlsruhe, Germany. Figure 1 (left) shows the *complementary cumulative distribution function* (CCDF) of the station degrees in a log-log plot. For a given value x, the CCDF describes the share of stations that have degree at least x. The CCDF closely resembles a straight line (in log-log scaling), indicating a *power-law distribution*. That means, there exists a real number β, the *power-law exponent*, such that the number of stations of degree x is roughly proportional to $x^{-\beta}$. We estimated the power-law exponents using the python package powerlaw [2]. For kvv, the exponent β is approximately 3.5. The goodness of fit is measured by the *Kolmogorov–Smirnov distance* (KS distance), which is the maximum absolute difference between the CCDFs of the measurement and of the assumed distribution. The KS distance for the kvv is 0.05. Table 1 reports both the power-law exponents and the corresponding KS distances. The estimated values of β, excluding the outlier vbb, indicate a high level of heterogeneity. As a side note, the power-law exponent for vbb is 4.1 with a KS distance of 0.03 when considering the whole network instead of the largest component. In contrast, the connection degrees are rather homogeneous, cf. e.g. kvv in Fig. 1 (right). A possible explanation is that long-distance trains stop less frequently.

Locality. To measure locality, we adapt the *bipartite clustering coefficient* [15]. Intuitively, it states how likely it is that two stations which share a connection are also contained together in a different connection, or that two connections containing the same station also have another station in common. For a formal definition, first note that we can interpret a HITTING SET instance (S, C) as a bipartite graph with the two partitions S and C and an edge joining $s \in S$ and $c \in C$ iff $s \in c$. Let $\#_{P_3}$ denote the number of paths of length 3 and $\#_{C_4}$ the number of cycles of length 4 in this graph. The bipartite clustering coefficient κ then is defined as $\kappa = 4 \cdot \#_{C_4}/\#_{P_3}$. It is the probability that a uniformly chosen 3-path is contained in a 4-cycle. Before computing κ, we normalize the bipartite graph by reducing it to its 2-core, which removes any attached trees. In doing so, the measure becomes more robust for our purpose, as attached trees do not impact the difficulty of an instance (they get removed by the reduction rules) while they decrease the clustering coefficient.

The clustering coefficients are reported in Table 1. All instances have a clustering coefficient of at least 0.25, which indicates a high level of locality. A possible explanation are the underlying geographic positions of the stations, with nearby stations likely appearing in the same connection.

Degree of Reduction. We measure the effectiveness of the reduction rules using the *relative core complexity*. It is the percentage of stations that remain after exhaustively applying the preprocessing. Table 1 shows that the resulting relative core complexity is very low for all 12 instances. This is in line with the original findings of Weihe [20], who applied the reduction rules on a few select European train networks. Moreover, it generalizes these results to networks of different scales, from urban to national. On the other hand, the 2-core is typically not much smaller than the original instance. This shows that Proposition 1 cannot explain the effectiveness of the reduction rules, which supports our previous assessment that the graph-theoretic perspective is not sufficient.

Judging from Table 1, we believe that heterogeneity of the stations and high locality are the crucial properties rendering the preprocessing so effective. Notwithstanding, it is also worth noting that the reduction rules work well on all instances, including vbb which is not very heterogeneous. The clustering on the other hand is high for all instances, indicating that locality is more important. Also, the db and vbb outliers seem to show that the influence of the station-connection ratio and the average station degree is limited. Though looking at these 12 networks can provide clues to what features are most important, it is not sufficient to draw a clear picture. In the following, we thoroughly test the effect of different properties on the effectiveness of the reduction rules by generating instances with varying properties.

4   Analysis of Generated Instances

This section discusses the generation and analysis of artificial HITTING SET instances. First, we present our model of generation which is based on the *geometric inhomogeneous random graphs* [5]. It allows creating networks with varying degree of heterogeneity and locality. We then analyze these instances with respect to the degree of reduction.

4.1 The Generative Model

In the field of network science, it is generally accepted that vertex degrees in realistic networks are heterogeneous [18]. A power-law distribution can be explained, inter alia, by the preferential attachment mechanism [3]. Beyond the generation of heterogeneous instances, different models have been proposed to also account for locality. The latter models typically use some kind of underlying geometry. One of the earliest works in that direction is by Watts and Strogatz [19]. More recently, and closer to our aim, Papadopoulos et al. [14] introduced the concept of popularity vs. similarity, making the creation of edges more likely, the more popular and similar the connected vertices are. They also observed that these two dimensions are naturally covered by the hyperbolic geometry, leading to hyperbolic random graphs [11]. Bringmann, Keusch, and Lengler [5] generalized this concept to *geometric inhomogeneous random graphs* (GIRGs). There, each vertex has a geometric position and a weight. Vertices are then connected by edges depending on their weights and distances. Despite a plethora of models for generating graphs, we are not aware of models generating heterogeneous HITTING SET instances. The closest is arguably the work by Giráldez-Cru and Levy [8], who generate SAT instances using the popularity vs. similarity paradigm.

To generate HITTING SET instances with varying heterogeneity and locality, we formulate a randomized model based on GIRGs. Each station and connection has a weight representing its importance. Moreover, stations and connections are randomly placed in a geometric space. The distance between stations and connections then provides a measure of similarity. In the HITTING SET instance, some station s is a member of connection c with a probability proportional to the combined weights of s and c and inverse proportional to the distance between the vertices s and c. To make this more precise, let $w_S \colon S \to \mathbb{R}$ and $w_C \colon C \to \mathbb{R}$ be two weight functions; we omit the subscript when no ambiguity arises. For $s \in S$ and $c \in C$, let $\mathrm{dist}(s, c)$ denote the geometric distance between the corresponding vertices. Finally, fix two positive constants $a, T > 0$. Then, station s is contained in connection c with probability

$$P(s,c) = \min\left\{1, \ \left(a \cdot \frac{w(s)w(c)}{\mathrm{dist}(s,c)}\right)^{1/T}\right\}. \tag{1}$$

The parameter a governs the expected degree. The *temperature* T controls the influence of the geometry. For $T \to 0$ the method converges to a step model, where s is contained in c if and only if $\mathrm{dist}(s,c) \le aw(s)w(c)$. Larger temperatures soften this threshold, allowing $s \in c$ for larger distances, and $s \notin c$ for smaller distances, with a low probability. Thus, T influences the locality of the instance.

The remaining degrees of freedom are the choice of the underlying geometry and the weights. For the geometry, we use the unit circle. Positions for stations and connections are drawn uniformly at random from $[0, 1]$ and the distance between $x, y \in [0, 1]$ is $\min\{|x - y|, 1 - |x - y|\}$. This is arguably the simplest possible symmetric geometry.

To choose the weights properly, it is important to note that the resulting degrees are expected to be proportional to the weights [5]. Thus, we mimic the real-world instances by choosing uniform weights for the connections and power-law weights, with varying exponent β, for the stations. It is not hard to see that for $\beta \to \infty$, the latter converge to uniform weights as well. In summary, adjusting β controls the heterogeneity.

4.2   Evaluation

We generate artificial networks and measure the dependence of their relative core complexity on the heterogeneity and locality. The size of an instance has three components: the (original) complexity $|S|$, the station-connection ratio $|S|/|C|$, and the average station degree δ_S. Note that these values also determine the number $|C|$ of connections and average connection degree δ_C. From the model, we have the two parameters we are most interested in, the power-law exponent β and the temperature T. We also consider the limit case of uniform weights for all vertices; slightly abusing notation, we denote this by $\beta = \infty$.

For the main part of the experiments, we used $|S| = 2000$, $|S|/|C| = 10.0$, and $\delta_S = 2.0$, leaning on the respective properties for the real-world instances. We let T vary between 0 and 1 in increments of 0.05, and β between 2 and 5 in increments of 0.25. For each combination, we generated ten samples. In the following, we first validate the data. Then we examine the influence of heterogeneity and locality. Afterwards we test whether our findings still hold true for different station-connection ratios and station degrees.

Data Validation. There are two aspects to the data validation. First, the instances should approximately exhibit the properties we explicitly put in, i.e., the values of $|S|$, $|S|/|C|$, δ_S, and the power-law behavior. Second, the implicit properties should also be as expected. In our case, we have to verify that changing T actually has the desired effect on the bipartite clustering coefficient κ.

Concerning the complexity $|S|$, note that a sampled instance per se does not need to be connected. If it is not, we again only use the largest component. Thus, when generating an instance with 2000 stations, the resulting complexity is actually a bit smaller. There are typically many isolated stations due to the small average station degree, this is particularly true for small power-law exponents. However, the complexity of the largest component never dropped below 1000 and usually was between 1400 and 1700 provided that $\beta > 2.5$. The transition to the largest component mainly meant ignoring isolated stations. Thus, also the station-connection ratio $|S|/|C|$ decreases slightly. For $\beta > 3$ it was typically around 8 and always at least 7. For smaller β, it is never below 5.

Recall that the average station degree δ_S is controlled by parameter a in Eq. 1. It is a constant in the sense that it is independent of the considered station-connection pair. However, it does depend on other parameters of the model. As there is no closed formula to determine a from δ_S, we estimated it

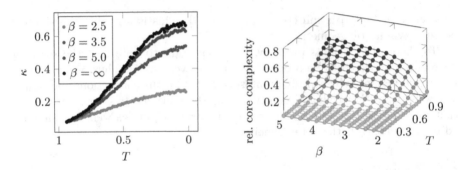

Fig. 2. (left) The clustering coefficient κ depending on the temperature T for different β. (right) The relative core complexity depending on the power-law exponent β and temperature T.

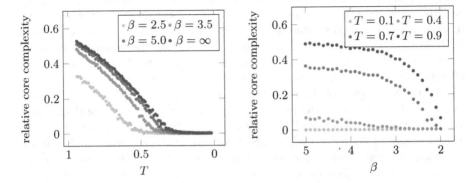

Fig. 3. The relative core complexity (left) depending on the temperature T (for different power-law exponents β), and (right) depending on β (for different T).

numerically. This estimation incurred some loss in accuracy but yielded values of δ_S between 1.9 and 2.1, very close to the desired $\delta_S = 2$. Transitioning to the largest component typically slightly increases δ_S, as the largest component is more likely to contain stations of higher degree. Anyway, δ_S never went above 2.7.

As with the real-world instances, we estimated the exponent β of the generated instances. For small values, the estimates matched the specified values. For larger exponents, the gap increases slightly, e.g., an estimated $\beta = 5.7$ for an instance with predefined parameter 5.0.

Finally, we examined the dependency between the bipartite clustering coefficient κ and the temperature T; see Fig. 2 (left). As desired, the clustering coefficient increases with falling temperatures. More precisely, κ ranges between 0 and some maximum attained at $T = 0$. The value of this maximum depends on the power-law exponent β, with smaller β giving smaller maxima.

Heterogeneity and Locality. For each instance described above, we computed the relative core complexity. To reduce noise, we look at the arithmetic mean of

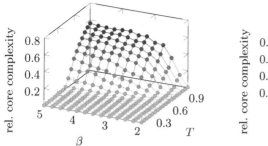 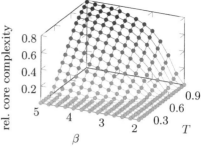

Fig. 4. The relative core complexity depending on the power-law exponent β and temperature T under alternative model parameters. (left) The station-connection ratio is $|S|/|C| = 4.0$ (instead of 10.0). (right) The average station degree is $\delta_S = 5.0$ (instead of 2.0).

ten samples for each parameter configuration. The measured core complexities and clustering coefficients in fact showed only small variance, with almost all values differing at most 5% from the respective mean. These measurements are presented in Fig. 2 (right), showing the mean core complexity depending on the temperature T and the power-law exponent β. For all parameter configurations, the relative core complexity was at most 50%. This is due to the low average station degree, which leads to dominant low-degree stations. More importantly, the core complexity varies strongly for different values of T and β.

The complexity decreased both with lower temperatures and lower power-law exponents. This further supports our claim that heterogeneity and locality both have a positive impact on the effectiveness of the reduction rules. The locality, however, seems to be more vital. To make this precise, low temperatures lead to small cores, independent of β, as shown in Fig. 3 (left). Even under uniform station weights ($\beta = \infty$), temperatures below 0.3 consistently produced instances with empty core. In contrast, the power-law exponent has only a minor impact. One can see in Fig. 3 (right) that for low temperatures, the core complexity is (almost) independent of β. For higher temperatures, the core complexities remain high over wide ranges of β, except for very low exponents.

In summary, high locality seems to be the most prominent feature that makes STATION COVER instances tractable, independent of their heterogeneity. Heterogeneity alone reduces the core complexity only slightly, except for extreme cases (very low power-law exponents). It is thus not the crucial factor. In the following, we verify this general behavior also for alternative model parameters such as station-connection ratio or average station degree.

Station-Connection Ratio. Recall that we fixed a ratio of $|S|/|C| = 10.0$ for the main part of our experiments. To see whether our observations are still valid for different settings, we additionally generated data sets with mostly the same parameters as before, except for $|S|/|C| = 4.0$. The result are shown in Fig. 4 (left). Comparing this to Fig. 2 (right), one can see that general depen-

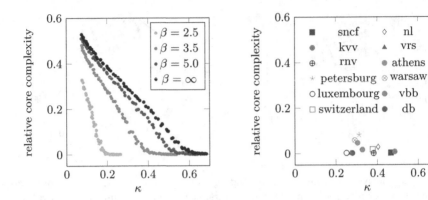

Fig. 5. The relative core complexity depending on the clustering coefficient κ (left) for generated instances with different values of the power-law exponent β, and (right) for the real-world instances.

dence on T and β is very similar. However, there are subtle differences. Under the smaller station-connection ratio the instances are tractable even for larger temperatures, up to $T = 0.5$ instead of the earlier 0.3, i.e., for lower locality. Also, the maximum core complexity over all combinations of T and β is larger than before, reaching almost 60%, compared to the 50% for $|S|/|C| = 10.0$. In summary, in the (more realistic) low-temperature regime, a lower station-connection ratio seems to further improve the effectiveness of the reduction rules.

Average Degree. To examine the influence of the average station degree δ_S, we generated another instance set with the same parameters as the main one, except that we increased the δ_S from 2.0 to 5.0. The results are shown in Fig. 4 (right). Again the general behavior is similar, but now lower temperatures are necessary to render the instances tractable. Moreover, the maximum core complexity increases significantly, reaching up to almost 100%. Generally, a smaller average degree makes the reduction rules more effective. Intuitively speaking, the existence of stations with low degree increases the likelihood that the reduction rule of station dominance can be applied.

Comparison with Real-World Instances. To compare generated and real-world instances more directly, we investigate the dependence between the relative core complexity and the bipartite clustering coefficient κ, instead of the model-specific temperature T. Figure 5 shows the results. Several of the real-world networks are well covered by the model. For example, our findings from the generated instances can be directly transferred to sncf and kvv. They have power-law exponents of 3.3 and 3.5, respectively, as well as clustering coefficients of at least 0.47, which explains their small core complexity. Moreover, for luxembourg, the model can explain the low complexity in spite of a small clustering coefficient of 0.25. The network exhibits a small exponent $\beta = 2.9$, which benefits the effectiveness of the preprocessing. On the other hand, petersburg has clustering $\kappa = 0.31$, but also a comparatively large core complexity of over 8%. Here, the main factor seems to be the high power-law exponent of 4.0.

Notwithstanding, there are also some real-world instances that have an unexpectedly low core complexity, which cannot be fully explained by the model. The vrs-instance has a low clustering coefficient of $\kappa = 0.27$ and a high power-law exponent of 3.5, but still a very low 0.1% core complexity. The reason seems to be its low average station degree of $\delta_S = 1.9$. The switzerland-instance also has a low core complexity of 1.7%, despite its high power-law exponent of $\beta = 4.5$ and a clustering coefficient of $\kappa = 0.33$. Especially the high value of β would point to a much higher complexity, however, its station-connection ratio $|S|/|C| = 5.6$ is significantly lower than that of the generated instances. All in all, the networks generated by the model are not perfectly realistic. However, the model does replicate properties that are crucial for the effectiveness of the reduction rules on real-world instances. Furthermore, in the interplay of heterogeneity and locality, it reveals locality as the more important property.

5   Impact on Other Domains

Although the focus of this paper is to understand which structural properties of public transportation networks make Weihe's reduction rules so effective, our findings go beyond that. Our experiments on the random model predict that HITTING SET instances in general can be solved efficiently if they exhibit high locality. Moreover, if the instance is highly heterogeneous, a smaller clustering coefficient suffices; see Fig. 5 (left). The element-set ratio and the average degree have, within reasonable bounds, only a minor impact on the effectiveness. Our experiments in Sect. 4.2 showed that instances are more difficult for a larger average degree and if the difference between the number of elements and the number of sets is high. Experiments not reported in this paper show that the latter is also true, if there are more sets than elements (i.e., the ratio is below 1).

Data Sets. We consider HITTING SET instances from three different applications; see Table 2. The first set of instances are metabolic reaction networks of Escherichia coli bacteria. The elements represent reactions and each set is a so-called *elementary mode*. Analyzing the hitting sets of these instances has applications in drug discovery. The corresponding data sets, labeled ec-*, were generated with the Metatool [10]. In the second type of instance, the sets consist of so-called *elementary pathways* that need to be hit by *interventions* that suppress all signals, which is relevant, inter alia, for the treatment of cancer. The data sets, EGFR.* and HER2.*, were obtained via the OCSANA tool [17]. The instances country-cover and language-cover are based on a country-language graph, taken from the network collection KONECT [12], with an edge between a country and a language if the language is spoken in that country. The corresponding HITTING SET instances ask for a minimum number of countries to visit to hear all languages, and for a minimum number of languages necessary to communicate with someone in every country, respectively.

Table 2. HITTING SET instances from other domains. Listed are the number $|S|$ of elements, the element-set ratio $|S|/|C|$, the bipartite clustering coefficient κ, and the relative core complexity.

Data set	$|S|$	$|S|/|C|$	κ	Core
ec-acetate	57	0.214	0.67	1.8%
ec-succinate	57	0.061	0.59	1.8%
ec-glycerol	60	0.028	0.66	1.7%
ec-glucose	58	0.009	0.66	36.2%
ec-combined	64	0.002	0.62	40.6%
EGFR.short	50	0.400	0.66	2.0%
EGFR.sub	56	0.239	0.57	1.8%
HER2.short	123	0.230	0.53	9.8%
HER2.sub	172	0.068	0.55	10.4%
country-cover	248	0.407	0.11	0.4%
language-cover	610	2.460	0.11	0.2%

Evaluation. The basic properties of the instances and the effectiveness of the reduction rules are reported in Table 2. The results match the prediction of our model: most instances have a high clustering coefficient and the reduction rules are very effective. The only instances that stand out at first glance are ec-glucose, ec-combined, HER2.short, and HER2.sub, which are not solved completely by the reduction rules despite their high clustering coefficients, as well as country-cover and language-cover, which are solved completely despite the comparatively low clustering coefficient of $\kappa = 0.11$.

However, a more detailed consideration reveals that these instances also match the predictions of the model. First, the two instances country-cover and language-cover are very heterogeneous with power-law exponent $\beta = 2.2$. As can be seen in Fig. 2 (left), a clustering coefficient of $\kappa = 0.11$ is already rather high for this exponent, leading to a low core complexity; see Fig. 5 (left).

The instances ec-glucose and ec-combined have skewed element-set ratios (more than 100 times as many sets as elements) and a high average degree (30 for the sets; 3k and 13k for the elements, respectively). Thus, these instances at least qualitatively match the predictions of the model that the reduction rules are less effective if the element-set ratio is skewed or the average degree is high. One obtains a similar but less pronounced picture for HER2.short and HER2.sub.

6   Conclusion

We explored the effectiveness of data reduction for STATION OVER on transportation networks. Our main finding is that real-world instances have high locality and heterogeneity, and that these properties make the reduction rules effective, with locality being the crucial factor. This directly transfers to general HITTING

SET instances. For future work, it would be interesting to rigorously prove that the reduction rules perform well on the model.

References

1. Abu-Khzam, F.N.: A kernelization algorithm for d-hitting set. J. Comput. Syst. Sci. **76**, 524–531 (2010)
2. Alstott, J., Bullmore, E., Plenz, D.: powerlaw: a Python package for analysis of heavy-tailed distributions. PloS One **9**, e85777 (2014)
3. Barabási, A.L., Albert, R.: Emergence of scaling in random networks. Science **286**, 509–512 (1999)
4. Bläsius, T., Fischbeck, P., Friedrich, T., Schirneck, M.: Understanding the effectiveness of data reduction in public transportation networks. arXiv:1905.12477 (2019)
5. Bringmann, K., Keusch, R., Lengler, J.: Sampling geometric inhomogeneous random graphs in linear time. In: Proceedings of the 25th Annual European Symposium on Algorithms (ESA), pp. 20:1–20:15 (2017)
6. Davies, J., Bacchus, F.: Solving MAXSAT by solving a sequence of simpler SAT instances. In: Lee, J. (ed.) CP 2011. LNCS, vol. 6876, pp. 225–239. Springer, Heidelberg (2011). https://doi.org/10.1007/978-3-642-23786-7_19
7. Gabaix, X.: Zipf's law for cities: an explanation. Q. J. Econ. **114**, 739–767 (1999)
8. Giráldez-Cru, J., Levy, J.: Locality in random SAT instances. In: Proceedings of the 26th International Joint Conference on Artificial Intelligence (IJCAI), pp. 638–644 (2017)
9. Jansen, B.M.P.: On structural parameterizations of hitting set: hitting paths in graphs using 2-SAT. J. Graph Algorithms Appl. **21**, 219–243 (2017)
10. von Kamp, A., Schuster, S.: Metatool 5.0: fast and flexible elementary modes analysis. Bioinformatics **22**, 1930–1931 (2006)
11. Krioukov, D., Papadopoulos, F., Kitsak, M., Vahdat, A., Boguná, M.: Hyperbolic geometry of complex networks. Phys. Rev. E **82**, 036106 (2010)
12. Kunegis, J.: KONECT - the Koblenz network collection. In: Proceedings of the 22nd International Conference on World Wide Web (WWW), pp. 1343–1350 (2013)
13. Niedermeier, R., Rossmanith, P.: An efficient fixed-parameter algorithm for 3-hitting set. J. Discrete Algorithms **1**, 89–102 (2003)
14. Papadopoulos, F., Kitsak, M., Serrano, M.Á., Boguñá, M., Krioukov, D.: Popularity versus similarity in growing networks. Nature **489**, 537–540 (2012)
15. Robins, G., Alexander, M.: Small worlds among interlocking directors: network structure and distance in bipartite graphs. Comput. Math. Organ. Theory **10**, 69–94 (2004)
16. Seidman, S.B.: Network structure and minimum degree. Soc. Netw. **5**, 269–287 (1983)
17. Vera-Licona, P., Bonnet, E., Barillot, E., Zinovyev, A.: OCSANA: optimal combinations of interventions from network analysis. Bioinformatics **29**, 1571–1573 (2013)
18. Voitalov, I., van der Hoorn, P., van der Hofstad, R., Krioukov, D.V.: Scale-free networks well done. arXiv:1811.02071 (2018)
19. Watts, D.J., Strogatz, S.H.: Collective dynamics of 'small-world' networks. Nature **393**, 440–442 (1998)
20. Weihe, K.: Covering trains by stations or the power of data reduction. In: Proceedings of the 1998 Algorithms and Experiments Conference (ALEX), pp. 1–8 (1998)

A Spatial Small-World Graph Arising from Activity-Based Reinforcement

Markus Heydenreich[1] and Christian Hirsch[2(✉)]

[1] Mathematisches Institut, Ludwig-Maximilians-Universität München,
Theresienstraße 39, 80333 München, Germany
m.heydenreich@lmu.de
[2] Institut für Mathematik, Universität Mannheim,
B6 26, 68161 Mannheim, Germany
hirsch@uni-mannheim.de

Abstract. In the classical preferential attachment model, links form instantly to newly arriving nodes and do not change over time. We propose a hierarchical random graph model in a spatial setting, where such a time-variability arises from an activity-based reinforcement mechanism. We show that the reinforcement mechanism converges, and prove rigorously that the resulting random graph exhibits the small-world property. A further motivation for this random graph stems from modeling synaptic plasticity.

Keywords: Random tree · Reinforcement · Neural network · Small-world graph

1 Introduction

Network Formation Driven by Reinforcement. Since the introduction of the *preferential attachment* model by Barabási and Albert [1], reinforcement mechanisms are recognized as a versatile tool in network formation. Why are preferential attachment models so popular? On the one hand, the resulting graphs exhibit universal features that are ubiquitous in real networks e.g., scale-free property, short distances [1,3]. In spatial versions of the preferential attachment mechanism, there is even strong local clustering [13]. A second reason for the popularity lies in the plausibility of the reinforcement scheme: When new agents enter the system, then they are more likely to link with highly connected agents than with those that have only few connections. The result is that a high degree is reinforced, some authors coin this the "Matthew effect" [21]. Even though the classical preferential attachment model is in principle a dynamical

This work is supported by The Danish Council for Independent Research—Natural Sciences, grant DFF – 7014-00074 *Statistics for point processes in space and beyond*, and by the *Centre for Stochastic Geometry and Advanced Bioimaging*, funded by grant 8721 from the Villum Foundation.

ⓒ Springer Nature Switzerland AG 2019
K. Avrachenkov et al. (Eds.): WAW 2019, LNCS 11631, pp. 102–114, 2019.
https://doi.org/10.1007/978-3-030-25070-6_8

model, the formation of edges occurs instantly, and is not changed with time, except for the addition of edges from new vertices. Variability in the formation of edges is thus not included in the classical preferential attachment model. Recent variants address this issue for instance by starting from a fixed number of nodes and then adding edges according to a preferential attachment mechanism [19].

Reinforcement effects are also typical for social sciences. Pemantle and Skyrms [18,20] study a mathematical model for a group of agents interacting with each other in such a way that every interaction makes the same interaction in the future more probable. Of particular interest is the long-term behavior: both on finite graphs [4] and on infinite networks [15] a nice characterization of the equilibrium states can be given: The reinforcement in the model is so strong that in the long run there is a formation of groups such that only the agents inside the groups interact but not across the groups. More precisely, it is shown that in an extremal equilibrium, the set of agents decomposes into finite sets, each of which includes a "center" that is always chosen by the other agents in that set.

Neural Networks. Reinforcement mechanisms are also typical for neural networks in the context of synaptic plasticity. To this end, we are considering a fairly simplistic model of a neural network: There is a set of neurons, each of them equipped with one axon and a number of dendrites which are connected to axons of other neurons. Pairs of axons and dendrites may form synapses, which are functional connections between neurons. However, not all geometric connections necessarily also form functional connections. The resulting network can be interpreted as a directed graph with neurons as vertices and synapses as edges (directed from dendrite to axon).

Experimental observation shows that the resulting neural network is rather sparse and very well connected, that is, any pair of neurons is connected through a short chain of neural connections reminiscent of the "small-world property". These features allow for very fast and efficient signal processing. The challenge is to explain the mechanism behind the formation of such sophisticated neural networks. Kalisman, Silberberg, and Markram [14] use experimental evidence to advocate a *tabula rasa approach* to the formation of these networks: In an early stage, there is a (theoretical) all-to-all geometrical connectivity. Stimulation and transmission of signals enhance certain touches to ultimately form functional connections, which results in a network with rather few actual synapses. This describes the plasticity of the brain at an early stage of the development.

A Mathematical Model. In order to model these effects mathematically, we consider a model of reinforced Pólya urns with graph-based competition. In this Pólya urn interpretation, the "color" of the balls in the Pólya urn represents the edges in a given graph (namely, the potential connections or touches).

This reinforcement scheme goes as follows: we start from a very large graph (e.g. the complete graph or a suitable grid), and initially equip all edges with weight one. The vertices are activated uniformly at random. If node v is activated at time $t \geq 0$, it queries the weight $W_{t_-}(v, w)$ of the edge (v, w) just *before* the activation. Then, the node chooses a neighbor w proportional to

$$W_{t-}(v,w)^\beta \qquad \beta > 0,$$

and the weight of the chosen edge increases by one. High weight of an edge thus means that this edge is chosen very often. We are interested in the subgraph formed by those edges whose weight is increasing linearly in time (i.e., edges that are chosen a positive fraction of time). Following the neural interpretation of the previous paragraph, these are the edges forming actual synapses.

The parameter $\beta > 0$ controls the strength of the reinforcement. We distinguish between *weak* reinforcement when $\beta < 1$ and *strong* reinforcement when $\beta > 1$. In the case of strong reinforcement, it is plausible that any stable equilibrium is concentrated on small "islands" which are not connected to each other; similar behavior is obtained rigorously for related models in [9,10,12]. On the other hand, if there is weak reinforcement, then all edges are contained in the limiting distribution, and thus no interesting subgraph is formed in the limit [5]. In the bordercase $\beta = 1$, there is linear reinforcement, where the classical Pólya urn (properly normalized) converges to a Dirichlet distribution. For our model of graph based interaction, the situation is more delicate, as it seems that the behavior for $\beta = 1$ resembles the subcritical regime [11].

We summarize that these conventional approaches yield interesting results, but they are not versatile enough to support the tabula rasa approach from a mathematical point of view: either the resulting functional connections form small local islands, or the entire network is kept in the limit. One might argue that our interpretation of neural interactions with reinforced Pólya urns is far too simplified, as there are more realistic mathematical models for brain activities (e.g. through a system of interacting Hawkes processes [6]). However, we do not expect that the overall picture as described above is changing by working in a more sophisticated setup. Instead, we are proposing a different route, where we introduce layers of neurons with varying fitness, and this leads indeed to an interesting network structure.

Mind that the reinforcement mechanism considered here models the plasticity of neurons, and should not be mistaken with reinforcement learning in the spirit of [16,22].

Our Contribution. In the present work, we are suggesting a new model for a network arising from reinforcement dynamics that are typical for the brain. Our model is built upon layers of spatial graphs, and the ability of neurons to form long connections. More precisely, in the base network of possible links, neurons at a higher layer have the potential of reaching further than neurons at lower layers, and a random fitness of neurons leads to a rapid coalescence of functional connections. We prove that the resulting graph is connected and cycle-free, and that far-away vertices are linked through a few edges only ("small-world"). In contrast to the preferential attachment model, which is based on reinforcement of degrees, our model reinforces edge activities.

Our main interest lies in the understanding of a versatile mathematical model for neural applications. It is clear that the actual formation of the brain involves much more complex processes that are beyond the scope of a rigorous treatment. Yet, we aim at clarifying which network characteristics can be explained by a simple reinforcement scheme, and which cannot.

The generality of our approach has the potential to be applied in a variety of contexts with different interpretations. Indeed, networks based on layered graphs are fundamental objects in machine learning, and therefore our model could contribute towards enhancing the understanding of "biologically plausible deep learning" in the spirit of [2].

Future Work. In the current model, the growing range of neurons at higher layers is defined externally. It appears desirable to extend the model such that this feature emerges from an intrinsic mechanism of self-organization. Additionally, one could envision replacing the externally defined fitnesses by a mechanism relying on the indegrees, thereby establishing a closer link to classical preferential attachment models. Even though in the present setting, we are deriving our results for layers of one-dimensional graphs, we expect that the main results also hold for higher-dimensional lattices. For example, when modelling the neurons in the visual pathway, layers of two-dimensional graphs seems more appropriate. Finally, it would be of interest to test the relevance of the proposed model with measurements in real world networks.

2   Model and Results

In this work, we consider a stochastic process of dynamically evolving edge weights $\{W_t(e)\}_{t \geq 0, e \in E}$ on the graph with nodes $V = \mathbb{Z} \times \mathbb{Z}_{\geq 0}$ and edges E given by pairs $((k, h), (\ell, h+1))$ for $|\ell - k| \leq a^h$ for some $a > 1$. Here, we think of $\mathbb{Z} \times \mathbb{Z}_{\geq 0}$ as an infinite number of layers, each consisting of infinitely many nodes. Additionally, the nodes feature iid heavy-tailed fitnesses $\{F_v\}_{v \in V}$ with tail index $\gamma < 1$. More precisely, we assume that $s^\gamma \mathbb{P}(F_v > s)$ remains bounded away from 0 and ∞ as $s \to \infty$. The assumption that the fitnesses are heavy tailed encourages that in higher layers we can observe nodes with outstandingly high fitness featuring a large number of ingoing edges from the previous layer.

At time $t = 0$, all edge weights are constant equal to 1, i.e., $W_0(e) = 1$ for every $e \in E$. To describe the evolution of $\{W_t\}_{t \geq 0}$, we equip the nodes of V with independent Poisson clocks. When at node $v = (k, h)$ the clock rings, then we choose one of the adjacent nodes w in the set $\mathcal{N}_v = \{(\ell, h+1) : |\ell - k| \leq a^h\}$ of out-neighbors and increment the weight of the incident edge by 1. According to the modeling paradigm described in Sect. 1, we prefer to choose fitter vertices and higher edge weights. More precisely, the probability to select $w = (\ell, h+1)$ is proportional to

$$F_w W_{t-}(v, w)^\beta,$$

where the parameter $\beta > 1$ describes the strength of the reinforcement bias. Note that when disregarding the fitnesses, the weight-evolutions of the outgoing edges at each node would form a classical Pólya urn with reinforcement parameter β. Moreover, conditioned on the fitnesses, activations at different nodes occur independently, so that one could also allow activation in bunches without influencing the distribution of the resulting random graph. Figure 1 illustrates the random graph model after a finite number of reinforcement steps. As will be made precise below, the apparent tree structure is no coincidence.

Having introduced the weight dynamics, we now extract the subgraph of relevant edges. More precisely, we let

$$\mathcal{E} = \{e \in E : \liminf_{t \to \infty} W_t(e)/t > 0\}$$

denote the subgraph consisting of edges that are reinforced a positive proportion of times. For the motivation from neutral networks, we may think of edges that are reinforced only a vanishing proportion of time as those potentially possible connections that after the self-organizing process of learning did not evolve into actual synapses.

The main result of this work establishes that $\mathcal{E}$ is a small-world graph in the sense that graph distances on $\mathcal{E}$ between layer-0 nodes grow logarithmically in their horizontal distance. To be more precise, by translation-invariance in the first coordinate, we may fix one of the vertices to be $(0,0)$ and therefore let H_N denote the graph distance on $\mathcal{E}$ between $(0,0)$ and $(N,0)$.

Theorem 1 (Typical distances; multiplicative version). *Let $a, \beta > 1$ and $\gamma < 1$. Then, asymptotically almost surely,*

$$\frac{H_N}{\log_a(N)} \xrightarrow{N \to \infty} 2.$$

Theorem 2 (Typical distances; additive version). *Let $a, \beta > 1$ and $\gamma < 1$. Then, there exists $c > 0$ such that for every $N, x \geq 1$*

$$\mathbb{P}(H_N \geq 2\log_a(N) + x) \leq \exp(-cx).$$

In particular, H_N is almost surely finite for every $N \geq 1$.

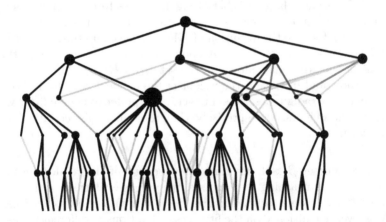

Fig. 1. Realization of the network model with parameters $a = 3$, $\beta = 3/2$ and $\gamma = 1/5$. Node diameters represent log-fitness values. Grayscales correspond to edge weights after 20 reinforcement steps.

Our results show that the activity-based reinforcement mechanism considered in our model is sufficient for constituting a sparse small-world graph. However, the heavy-tailed fitnesses and (externally defined) layers of growing range are vital ingredients.

The almost sure finiteness of H_N for all $N \geq 1$ means that all nodes at layer 0 are connected in $\mathcal{E}$. In fact, all other nodes are connected as well. Since $\mathcal{E}$ does not contain cycles (Lemma 1), it is therefore a tree.

Note that Theorems 1 and 2 can form the starting points for building a consistent estimator for the model parameter a. Once a is estimated, the parameter γ could be estimated from the indegree distribution. Finally, the precise choice of the parameter $\beta > 1$ does not influence the distribution of $\mathcal{E}$, so that its estimation would only be feasible on the basis of snapshots after finite time.

3   Proofs

First, in Sect. 3.1, we establish the lower bound of Theorem 1. The main step is to show that the relevant edges $\mathcal{E}$ form a forest. That is, with probability 1, every node has precisely one outgoing edge in $\mathcal{E}$. The argument critically relies on the assumption of strong reinforcement, where $\beta > 1$.

Next, the additive upper bound in Theorem 2 is stronger than the multiplicative upper bound in Theorem 1, so that it suffices to establish the former. To achieve this goal, in Sect. 3.2, we first give a short and instructive proof for $a \geq 3$. The heavy-tailedness of the fitness distribution ensures that although the number of possible connections from each node grows exponentially in the layer, with positive probability, $\mathcal{E}$ contains the edge leading to the node with maximal fitness in the next layer. Then, in Sect. 3.3, we work out the more subtle arguments for general $a > 1$.

3.1   Lower Bound

The main step in the lower bound is to prove that $\mathcal{E}$ is a forest. Essentially, this follows from a variant of the celebrated Rubin's theorem for Pólya urns in the regime of strong reinforcement.

Lemma 1 ($\mathcal{E}$ is a forest).   *With probability 1, $\mathcal{E}$ is a forest.*

Proof. The critical observation is that the outgoing edges adjacent to a node v are only reinforced at Poisson clock rings at the vertex v. Hence, we may view these edges as colors in a Pólya urn governed by a super-linear reinforcement scheme. Then, by the celebrated Rubin's theorem, almost surely all but one of the edges are reinforced only finitely often. For two colors, this is shown in [17, Theorem 3.6], and a generalization to an arbitrary finite number can be found in [23, Theorem 3.3.1].

Hence, in each layer $h \geq 0$, there exists almost surely a unique node (L_h, h) such that $(0,0)$ connects to (L_h, h) by a directed path in $\mathcal{E}$. Similarly, we write

(R_h, h) for the unique node in layer h connected along a directed path to $(N, 0)$. In this notation, H_N is twice the coalescence time of L_h and R_h, i.e.,

$$H_N = 2\inf\{h \geq 1 : L_h = R_h\}. \tag{1}$$

Now, the lower bound becomes a consequence of the structure of the underlying deterministic graph (V, E). More precisely, we first establish an auxiliary result on the growth of the difference $D_h = R_h - L_h$.

Lemma 2 (Growth of D_h). *Let $h, h' \geq 0$. Then,*

$$|D_{h+h'} - D_h| \leq 2 \sum_{h \leq i < h+h'} a^i \leq \frac{2a^{h+h'}}{a-1}.$$

Proof. By definition, any node in layer i can connect to nodes in layer $i+1$ at horizontal distance at most a^i, so that for every $i \geq 0$,

$$\max\{|L_i - L_{i+1}|, |R_i - R_{i+1}|\} \leq a^i.$$

In particular,

$$|D_{h+h'} - D_h| \leq \sum_{h \leq i < h+h'} |D_{i+1} - D_i| \leq 2 \sum_{h \leq i < h+h'} a^i$$

The second inequality in the assertion follows from the geometric series representation. $\qquad\square$

Now, we have all ingredients to prove the lower bound in Theorem 1.

Proof (Proof of Theorem 1, lower bound). Since $D_{H_N/2} = 0$ and $D_0 = N$, an application of Lemma 2 gives that

$$\log_a(2/(a-1)) + H_N/2 \geq \log_a(N),$$

as asserted. $\qquad\square$

3.2   Theorem 2; $a \geq 3$

Lemma 1 produces for every node a unique outgoing edge that is reinforced infinitely often. Leveraging the heavy-tailedness of the fitness distribution, a key ingredient in the proof of the upper bound is that with a probability bounded away from 0, this edge leads to the node with maximal fitness. To make this precise, let $v^{\max} \in \mathcal{N}_v$ be the out-neighbor of $v \in V$ with maximal fitness. In later parts of the manuscript, we also use $v^{\max}$ as a generic notation, when $v \in V$ is a specific node. We let

$$\mathcal{F}_h = \sigma(\{L_{h'}, R_{h'}\}_{h' \leq h}, \{F_v\}_{v \in \mathbb{Z} \times \{0,...,h\}})$$

denote the σ-algebra of the information gathered up to layer h and write

$$\mathcal{F}_h^* = \sigma(\{L_{h'}, R_{h'}\}_{h' \leq h}, \{F_v\}_{v \in \mathbb{Z} \times \{0,...,h+1\}})$$

for the σ-algebra that additionally contains the information on the fitnesses in layer $h+1$.

Lemma 3 (Choice of the fittest). *For every $\varepsilon > 0$ there exists $q_\varepsilon > 0$ such that almost surely for every $h \geq 0$ and $v \in \mathbb{Z} \times \{h\}$,*

$$\mathbb{P}(\{v, v^{\mathsf{max}}\} \in \mathcal{E} \mid \mathcal{F}_h^*) \geq q_\varepsilon \mathbb{1}\{E_v^{\mathsf{max-fit}}\},$$

where $E_v^{\mathsf{max-fit}} = \{\max_{w \in \mathcal{N}_v} F_w \geq \varepsilon \sum_{w \in \mathcal{N}_v} F_w\}$.

Proof. Let τ_n denote the nth firing time at the node v and write

$$E_v^n = \{W_{\tau_n}(\{v, v^{\mathsf{max}}\}) = W_{\tau_n-}(\{v, v^{\mathsf{max}}\}) + 1\}$$

for the event that at time τ_n the edge $\{v, v^{\mathsf{max}}\}$ is reinforced. In particular, $\{\{v, v^{\mathsf{max}}\} \in \mathcal{E}\} \supset \cap_{n \geq 1} E_v^n$ and

$$\mathbb{P}\left(E_v^n \,\middle|\, \mathcal{F}_h^*, \bigcap_{k \leq n-1} E_v^k\right) \geq \frac{F_{v^{\mathsf{max}}} n^\beta}{F_{v^{\mathsf{max}}} n^\beta + \sum_{w \in \mathcal{N}_v \setminus \{v^{\mathsf{max}}\}} F_w}.$$

Therefore, putting $M_v = \max_{w \in \mathcal{N}_v} F_w$ and $S_v = \sum_{w \in \mathcal{N}_v} F_w$, we obtain that almost surely,

$$\mathbb{P}(\{v, v^{\mathsf{max}}\} \in \mathcal{E} \mid \mathcal{F}_h^*) \geq \prod_{n \geq 1} \frac{M_v n^\beta}{M_v n^\beta + S_v}.$$

In particular,

$$\mathbb{P}(\{v, v^{\mathsf{max}}\} \in \mathcal{E} \mid \mathcal{F}_h^*) \geq \mathbb{1}\{M_v \geq \varepsilon S_v\} \prod_{n \geq 1} (1 - (1 + \varepsilon n^\beta)^{-1}).$$

Since the series $\sum_{n \geq 1} (1 + \varepsilon n^\beta)^{-1}$ converges, the product $q_\varepsilon = \prod_{n \geq 1} (1 - (1 + \varepsilon n^\beta)^{-1})$ is strictly positive, as asserted.

To show that the sum and the maximum of the fitnesses appearing in Lemma 3 are of the same order, we critically rely on the assumption that the fitnesses are heavy-tailed. In order to be applicable both for the case $a \geq 3$ as well as for $a < 3$, we provide a slightly more refined result, where we compare the second largest value of iid heavy-tailed Pareto random variables to the sum. For this purpose, we write $\max_{i \leq m}^{(2)} x_i$ for the second largest value among real numbers $x_1, \ldots, x_m$.

Lemma 4 (Sum vs. Second-largest value for Pareto random variables). *Let $\{X_i\}_{i \geq 1}$ be iid Pareto random variables with parameter $\gamma < 1$. Let $S_m = \sum_{i \leq m} X_i$ and let $M_m^{(2)} = \max_{i \leq m}^{(2)} X_i$. Then, $\{S_m / M_m^{(2)}\}_{m \geq 1}$ is tight.*

Mind that Lemma 4 implies readily that $\{S_m / \max_{i \leq m} X_i\}_{m \geq 1}$ is tight as well.

Proof. First, by the stable limit theorem, the scaled sum $m^{-1/\gamma} S_m$ converges to a stable distribution [8, Theorem XVII.5.3]. Second, by extremal value theory, the scaled maximum $m^{-1/\gamma} \max_{i \leq m} X_i$ converges to a Fréchet distribution, whereas the ratio $\max_{i \leq m} X_i / M_m^{(2)}$ converges to 1 [7, Theorem 3.3.7, Example 4.1.11]. This yields tightness of $S_m / M_m^{(2)}$.

Now, we prove Theorem 2 for $a \geq 3$. The key simplification in the case $a \geq 3$ is that for every $h \geq 0$, the set of possible coalescence nodes grows so quickly that the conditional probability of coalescence in step $h+1$ given the information in $\mathcal{F}_h$ is bounded away from 0 uniformly in $h \geq 0$.

Proof. (Proof of Theorem 2, $a \geq 3$). To prove the result, we first assert that there exists $\delta > 0$ such that

$$\mathbb{P}(L_{h+1} = R_{h+1} \mid \mathcal{F}_h) \geq \delta$$

holds for every $h \geq \log_a(N) + \log_a(2)$. Once this assertion is shown, we obtain that for $x \geq 2\log_a(2)$,

$$\mathbb{P}(H_N \geq 2\log_a(N) + x) \leq \mathbb{P}(L_h \neq R_h \text{ for all } h \leq \log_a(N) + x/2)$$
$$\leq (1 - \delta)^{\lfloor x/2 - \log_a(2) \rfloor},$$

which decays exponentially fast in x.

To prove the asserted lower bound, we first introduce

$$C_h^+ = \mathcal{N}_{(L_h, h)} \cup \mathcal{N}_{(R_h, h)} \quad \text{and} \quad C_h^- = \mathcal{N}_{(L_h, h)} \cap \mathcal{N}_{(R_h, h)} \tag{2}$$

as the union and intersection of the out-neighborhoods of (L_h, h) and (R_h, h), respectively. Then,

$$\{L_{h+1} = R_{h+1}\} \supset \{L_h^{\max} = R_h^{\max}\} \cap A_h,$$

where

$$A_h = \{L_{h+1} = L_h^{\max}\} \cap \{R_{h+1} = R_h^{\max}\}.$$

Therefore, by Lemma 3, for every $\varepsilon > 0$,

$$\mathbb{P}(L_{h+1} = R_{h+1} \mid \mathcal{F}_h)$$
$$\geq \mathbb{E}\left[\mathbb{1}\{E_h^{\mathsf{LR}}\}\mathbb{P}(A_h \mid \mathcal{F}_h^*) \mid \mathcal{F}_h\right]$$
$$\geq q_\varepsilon^2 \mathbb{P}\left(\{L_h^{\max} = R_h^{\max}\} \cap \left\{\max_{w \in C_h^+} F_w \geq \varepsilon \sum_{w \in C_h^+} F_w\right\} \mid \mathcal{F}_h\right),$$

where $E_h^{\mathsf{LR}} = \{L_h^{\max} = R_h^{\max}\}$. Since the positions $L_h^{\max}$, $R_h^{\max}$ of the maximal fitnesses are independent of the value of the sum and the value of the maximum of the relevant fitnesses, we arrive at

$$\mathbb{P}(L_{h+1} = R_{h+1} \mid \mathcal{F}_h) \geq q_\varepsilon^2 \mathbb{P}(L_h^{\max} = R_h^{\max} \mid \mathcal{F}_h)\mathbb{P}\left(\left\{\max_{w \in C_h^+} F_w \geq \varepsilon \sum_{w \in C_h^+} F_w\right\} \mid \mathcal{F}_h\right).$$

By Lemma 4, the second probability is bounded below by $1/2$ for sufficiently small $\varepsilon > 0$. Hence, it remains to provide a lower bound for $\mathbb{P}(L_h^{\max} = R_h^{\max} \mid \mathcal{F}_h)$.

We write $P_h \in \mathbb{Z} \times \{h+1\}$ for the position of the maximal fitness in C_h^+, i.e.,

$$F_{P_h} = \max_{w \in C_h^+} F_w.$$

Then, P_h is uniformly distributed in C_h^+, so that

$$\mathbb{P}(L_h^{\max} = R_h^{\max} \mid \mathcal{F}_h) = \mathbb{P}(P_h \in C_h^- \mid \mathcal{F}_h) = \frac{\#C_h^-}{\#C_h^+} \geq \frac{\#C_h^-}{4\lfloor a^h \rfloor + 2}.$$

Finally, to derive a lower bound on $\#C_h^-$, Lemma 2 gives that, for every $h \geq \log_a(N) + \log_a(2)$,

$$|L_h - R_h| \leq N + \frac{2a^h}{a-1} \leq N + a^h \leq \frac{3}{2}a^h.$$

Therefore,

$$\#C_h^- \geq 2\lfloor a^h \rfloor - \frac{3}{2}a^h \geq \frac{1}{4}a^h,$$

which implies the required lower bound.

3.3 Theorem 2; $a < 3$

After having developed the intuition behind the proof of Theorem 2 for $a \geq 3$, we now assume that $a < 3$. The arguments in this case are more involved since it may happen that L_h and R_h are so far away that the set C_h^- of possible coalescence points from (2) becomes empty. We deal with this problem by establishing that L_h and R_h both do not move substantially for a finite number of steps, which guarantees that the set of possible coalescence points becomes non-empty again.

We start by showing that coalescence occurs with positive probability after a small number of steps if initially L_h and R_h are not too far apart.

Lemma 5 (H_N is small with positive probability). *There exists $k \geq 1$ such that*

$$\inf_{\substack{h,N \geq 0 \\ z:\,|z| \leq 4a^h/(a-1)}} \mathbb{P}(L_{h'} = R_{h'} \text{ for some } h' \in [h, h+k-1] \mid L_h - R_h = z) > 0.$$

Proof. First, if $|z| \leq a^h$, then $\#C_h^- \geq a^h/2$ for large $h \geq 0$, so that arguing as in Sect. 3.2 yields that

$$\mathbb{P}(L_{h+1} = R_{h+1} \mid L_h - R_h = z) \geq \delta_0$$

for a sufficiently small value of $\delta_0 > 0$.

Hence, we may assume that $|z| > a^h$ and introduce the events

$$E_{h'} = \{L_{h'+1} = R_{h'+1}\} \cup \{\max\{|L_{h'+1} - L_{h'}|, |R_{h'+1} - R_{h'}|\} \leq \varepsilon_1 a^{h'}\}, \quad (3)$$

where $\varepsilon_1 = (a-1)/8$. We assert that there exists $\delta > 0$ such that for every $h' \geq h$

$$\mathbb{P}(E_{h'} \mid \mathcal{F}_{h'}) \geq \delta. \quad (4)$$

Before proving (4), we show how to conclude the proof of the lemma. First, set

$$h_1 = \min \left\{ h' \geq h : |z| + 2\varepsilon_1 \sum_{h \leq i \leq h'-1} a^i \leq \frac{1}{2} a^{h'} \right\}.$$

In particular,

$$|z| \geq \frac{a^{h_1-1}}{2} - 2\varepsilon_1 \sum_{h \leq i \leq h_1-2} a^i = \frac{a^{h_1-1}}{2} - \frac{2\varepsilon_1(a^{h_1-1} - a^h)}{a-1} \geq \frac{a^{h_1-1}}{4}.$$

Then, $|z| \leq 4a^h/(a-1)$ implies that $h_1 - h - 1 \leq \log_a(16/(a-1))$. Note that if $E_{h'} \cap \{L_{h'} \neq R_{h'}\}$ occurs for every $h' \in [h, \ldots, h_1 - 1]$, then

$$|D_{h_1}| \leq \left| |D_{h_1}| - |z| \right| + |z| \leq \frac{2\varepsilon_1 a^{h_1}}{a-1} + \frac{a^{h_1}}{2} \leq a^{h_1},$$

where in the second inequality, we insert the definition of h_1 to bound $|z|$. Hence, by the case considered at the beginning of the proof, the conditional probability that $L_{h_1+1} = R_{h_1+1}$ given that $\cap_{h \leq h' < h_1} E_{h'}$ is bounded below by δ_0. Taking everything together, we arrive at the asserted positive lower bound

$$\mathbb{P}\left(\{L_{h_1+1} = R_{h_1+1}\} \cap \bigcap_{h \leq h' < h_1} E_{h'} \,\middle|\, L_h - R_h = z \right) \geq \delta_0 \delta^{h_1-h} \geq \delta_0 \delta^{\log_a(16/(a-1))}.$$

It remains to establish (4). To that end, we first set as before

$$A_{h'} = \{L_{h'+1} = L_{h'}^{\max}\} \cap \{R_{h'+1} = R_{h'}^{\max}\}.$$

Then, we let $P_{h'} = (p_{h'}, h'+1)$ and $P_{h'}^{(2)} = (p_{h'}^{(2)}, h'+1)$ denote the positions of the largest and the second largest fitness in the union set $C_{h'}^+$ as defined in (2). That is,

$$F_{P_{h'}} = \max_{w \in C_{h'}^+} F_w \quad \text{and} \quad F_{P_{h'}^{(2)}} = \max_{w \in C_{h'}^+}{}^{(2)} F_w.$$

Now, define

$$E'_{h'} = \left\{ \max\{|p_{h'} - L_{h'}|, |p_{h'}^{(2)} - R_{h'}|\} \leq \varepsilon_1 a^{h'} \right\}$$

as the event that the distances between $p_{h'}$ and $L_{h'}$, as well as between $p_{h'}^{(2)}$ and $R_{h'}$ are at most $\varepsilon_1 a^{h'}$. Then, we claim that

$$E_{h'} \supset E'_{h'} \cap A_{h'}.$$

Indeed, assume that $E'_{h'}$ occurs, so that from $\varepsilon_1 < 1$ we obtain that $p_{h'}^{(2)} \in \mathcal{N}_{R_{h'}}$. In that case, if $P_{h'}$ is contained in the intersection set $C_{h'}^-$, then $L_{h'}^{\max} = R_{h'}^{\max}$. Otherwise, $p_{h'} = L_{h'}^{\max}$ and $p_{h'}^{(2)} = R_{h'}^{\max}$, so that $\max\{|L_{h'}^{\max} - L_{h'}|, |R_{h'}^{\max} - R_{h'}|\} \leq \varepsilon_1 a^{h'}$. In particular,

$$\{L_{h'}^{\max} = R_{h'}^{\max}\} \cup \{\max\{|L_{h'}^{\max} - L_{h'}|, |R_{h'}^{\max} - R_{h'}|\} \leq \varepsilon_1 a^{h'}\}$$

occurs. Hence, under $A_{h'}$ the previous line becomes the defining equation for $E_{h'}$ as in (3).

Now, arguing as in the case $a \geq 3$, we derive that for every $\varepsilon > 0$,

$$\mathbb{P}(E_{h'} \mid \mathcal{F}_h) \geq q_\varepsilon^2 \mathbb{P}(E'_{h'} \mid \mathcal{F}_h) \mathbb{P}\left(\left\{ \max_{w \in C_h^+}{}^{(2)} F_w \geq \varepsilon \sum_{w \in C_h^+} F_w \right\} \middle| \mathcal{F}_h\right).$$

Note that here, we need to consider the second largest value in C_h^+ since under $E'_{h'}$ the positions $L_{h'}^{\max}$ and $R_{h'}^{\max}$ could be distinct. By Lemma 4, it therefore suffices to derive a lower bound on $\mathbb{P}(E'_{h'} \mid \mathcal{F}_h)$.

Finally, since p_h and $p_{h'}^{(2)}$ are uniform in $C_{h'}^+$, we obtain that for large $h \geq 0$

$$\mathbb{P}\left(\max\{|p_{h'} - L_{h'}|, |p_{h'}^{(2)} - R_{h'}|\} \leq \varepsilon_1 a^{h'} \mid \mathcal{F}_{h'} \right) \geq \left(\frac{\varepsilon_1 a^{h'}}{4a^{h'} + 2} \right)^2,$$

which is bounded away from 0, thereby completing the proof of (4).

Proof (Proof of Theorem 2, $a < 3$). First, using $1 \leq 2/(a-1)$ for $a < 3$, Lemma 2 implies that for every $h \geq h_0 = \log_a(N)$,

$$|D_h| \leq N + \frac{2a^h}{a-1} \leq \frac{4a^h}{a-1},$$

so that we can apply Lemma 5. With $k \geq 1$ as in that lemma, we let

$$G_i = \{R_h \neq L_h \text{ for every } h \in [ik, i(k+1) - 1]\}$$

denote the event that we do not see coalescence in the interval $[ik, i(k+1) - 1]$ and put $G'_i = \cap_{i' \leq i} G_{i'}$. In particular, under the event $\{H_N \geq 2\log_a(N) + x\}$, the event G'_{i_1} occurs for $i_1 = \lfloor (\log_a(N) + x/2)/k \rfloor$. Hence, by the Markov property at time $(i_1 - 1)$ and Lemma 5, we have a constant $\varepsilon > 0$ such that

$$\mathbb{P}(G'_{i_1}) \leq \mathbb{E}\left[\mathbb{P}\left(L_h \neq R_h \text{ for every } h \in [(i_1 - 1)k, i_1 k - 1] \mid D_{(i_1-1)k}\right) \mathbb{1}\{G'_{i_1-1}\} \right]$$
$$\leq (1 - \varepsilon)\mathbb{P}(G'_{i_1-1}).$$

Hence, putting $i_0 = \lceil \log_a(N)/k \rceil$, we conclude that

$$\mathbb{P}(H_N \geq 2\log_a(N) + x) \leq (1 - \varepsilon)^{i_1 - i_0},$$

which decays exponentially fast in x.

Acknowledgments. The authors thank all anonymous referees. We also thank C. Leibold for interesting discussions on the neuro-scientific background of synaptic plasticity and comments on an earlier version of the manuscript.

References

1. Barabási, A.L., Albert, R.: Emergence of scaling in random networks. Science **286**, 509–512 (1999)
2. Bengio, Y., Lee, D., Bornschein, J., Lin, Z.: Towards biologically plausible deep learning. CoRR abs/1502.04156 (2015). http://arxiv.org/abs/1502.04156
3. Bollobás, B., Riordan, O.: The diameter of a scale-free random graph. Combinatorica **24**(1), 5–34 (2004)
4. Bonacich, P., Liggett, T.M.: Asymptotics of a matrix valued Markov chain arising in sociology. Stoch. Process. Appl. **104**(1), 155–171 (2003)
5. Couzinié, Y., Hirsch, C.: Infinite WARM graphs I. Weak reinforcement regime (in preparation)
6. Delattre, S., Fournier, N., Hoffmann, M.: Hawkes processes on large networks. Ann. Appl. Probab. **26**(1), 216–261 (2016)
7. Embrechts, P., Klüppelberg, C., Mikosch, T.: Modelling Extremal Events. Springer, Berlin (1997). https://doi.org/10.1007/978-3-642-33483-2
8. Feller, W.: An Introduction to Probability Theory and its Applications, vol. II, 2nd edn. Wiley, New York (1971)
9. Hirsch, C., Holmes, M., Kleptsyn, V.: Absence of WARM percolation in the very strong reinforcement regime, preprint available at https://christian-hirsch.github.io/publications.html
10. Van Der Hofstad, R., Holmes, M., Kuznetsov, A., Ruszel, W.: Strongly reinforced Pólya urns with graph-based competition. Ann. Appl. Probab. **26**(4), 2494–2539 (2016)
11. Holmes, M., Kleptsyn, V.: Infinite WARM graphs. Critical regime (in preparation)
12. Holmes, M., Kleptsyn, V.: Proof of the WARM whisker conjecture for neuronal connections. Chaos **27**(4), 043104 (2017)
13. Jacob, E., Mörters, P.: A spatial preferential attachment model with local clustering. In: Bonato, A., Mitzenmacher, M., Prałat, P. (eds.) WAW 2013. LNCS, vol. 8305, pp. 14–25. Springer, Cham (2013). https://doi.org/10.1007/978-3-319-03536-9_2
14. Kalisman, N., Silberberg, G., Markram, H.: The neocortical microcircuit as a tabula rasa. Proc. Natl. Acad. Sci. **102**(3), 880–885 (2005)
15. Liggett, T.M., Rolles, S.W.W.: An infinite stochastic model of social network formation. Stoch. Process. Appl. **113**(1), 65–80 (2004)
16. Montague, P.R., Dayan, P., Sejnowski, T.J.: A framework for mesencephalic dopamine systems based on predictive Hebbian learning. J. Neurosci. **16**(5), 1936–1947 (1996)
17. Pemantle, R.: A survey of random processes with reinforcement. Probab. Surv. **4**, 1–79 (2007)
18. Pemantle, R., Skyrms, B.: Network formation by reinforcement learning: the long and medium run. Math. Soc. Sci. **48**(3), 315–327 (2004)
19. Pittel, B.: On a random graph evolving by degrees. Adv. Math. **223**(2), 619–671 (2010)
20. Skyrms, B., Pemantle, R.: A dynamic model of social network formation. Proc. Natl. Acad. Sci. USA **97**(16), 9340–9346 (2000)
21. Stanovich, K.E.: Matthew effects in reading: some consequences of individual differences in the acquisition of literacy. J. Educ. **189**(1–2), 23–55 (2009)
22. Sutton, R.S., Barto, A.G.: Reinforcement Learning: An Introduction, 2nd edn. MIT Press, Cambridge (2018)
23. Zhu, T.: Nonlinear Pólya urn models and self-organizing processes. Ph.D. thesis, University of Pennsylvania (2009)

SimpleHypergraphs.jl—Novel Software Framework for Modelling and Analysis of Hypergraphs

Alessia Antelmi[1], Gennaro Cordasco[2], Bogumił Kamiński[3], Paweł Prałat[4], Vittorio Scarano[1], Carmine Spagnuolo[1], and Przemyslaw Szufel[3(✉)]

[1] Dipartimento di Informatica, Università degli Studi di Salerno, Fisciano, Italy
{aantelmi,vitsca,cspagnuolo}@unisa.it
[2] Dipartimento di Psicologia, Università degli Studi della Campania
"Luigi Vanvitelli", Caserta, Italy
gennaro.cordasco@unicampania.it
[3] SGH Warsaw School of Economics, Warsaw, Poland
{bkamins,pszufe}@sgh.waw.pl
[4] Department of Mathematics, Ryerson University, Toronto, ON, Canada
pralat@ryerson.ca

Abstract. Hypergraphs are natural generalization of graphs in which a single (hyper)edge can connect any number of vertices. As a result, hypergraphs are suitable and useful to model many important networks and processes. Typical applications are related to social data analysis and include situations such as exchanging emails with several recipients, reviewing products on social platforms, or analyzing security vulnerabilities of information networks. In many situations, using hypergraphs instead of classical graphs allows us to better capture and analyze dependencies within the network. In this paper, we propose a new library, named `SimpleHypergraphs.jl`, designed for efficient hypegraph analysis. The library exploits the Julia language flexibility and direct support for distributed computing in order to bring a new quality for simulating and analyzing processes represented as hypergraphs. In order to show how the library can be used we study two case studies based on the Yelp dataset. Results are promising and confirm the ability of hypergraphs to provide more insight than standard graph-based approaches.

Keywords: Hypergraphs · Modelling hypergraphs · Software library · Julia programming language

1 Introduction

Many human-technology interaction situations generate data that can be viewed, based on the type of interaction, as a self-organizing network. In these networks

The research is financed by NAWA—The Polish National Agency for Academic Exchange.

© Springer Nature Switzerland AG 2019
K. Avrachenkov et al. (Eds.): WAW 2019, LNCS 11631, pp. 115–129, 2019.
https://doi.org/10.1007/978-3-030-25070-6_9

(for example, the Yelp on-line social network) nodes not only contain some useful information (such as user's profile, photos, reviews) but are also internally connected to other nodes (relations based on similar user's behaviour, similar taste, age, geographic location). Indeed, the proliferation of cellular usage has given rise to massive amounts of data that, through data mining and analytics, promises to reveal a wealth of information on how users interact with one another and shape the preferences of others.

Hypergraphs are of particular interest in the field of knowledge discovery where most problems currently modelled as graphs would be more accurately modelled as hypergraphs. Indeed, hypergraphs are natural generalization of graphs where one edge consists of several vertices instead of just a pair of vertices. This feature makes the hypergraphs particularly useful for modeling real world systems in which many references occur simultaneously. Examples include sending emails to many people, co-authorships of scientific publications, or several parties participating in a crypto-currency transaction. All of these complex real-world systems can be efficiently modelled with hypergraphs. Moreover, hypergraphs can also be extremely helpful, in computational social science, for the development of computer simulations [18,19]. Indeed, hypergraphs can be used to model any complex interaction among a group of simulated agents. Despite this fact, the theory and tools are not sufficiently developed to allow most problems to be tackled directly within this context.

The goal of this paper is to introduce a new library designed for efficient hypergraph analysis in the Julia language named `SimpleHypergraphs.jl`. The library makes an excessive use and is built on top of `LightGraphs.jl`, which is an efficient high-performance engine for graph analytics. Combined with language flexibility and direct support for distributed computing, the `SimpleHypergraphs.jl` library can bring a new quality for simulating and analyzing processes represented as hypergraphs.

The paper is structured as follows. In Sect. 2, we start with a review of the existing frameworks dealing with hypergraphs, describe the motivation and introduce the `SimpleHypergraphs.jl` library and its functionality. In Sect. 3, a use case with analysis of Yelp reviews is presented with the aim to show a real-life application of the developed library. Finally, we sum up the paper in the Conclusions section.

2   Modelling and Hypergraphs with `SimpleHypergraphs.jl`

In this section we start by introducing motivation for the hypegraph library and next we move towards describing its functionality.

2.1   Motivation

Despite the fact that hypergraphs are natural representations of many real-world systems, there are currently very few software frameworks that are suitable for modelling and mining hypergraphs. In this Section, we give a brief overview of several software libraries, focusing on their code availability and capability in modelling and analyzing hypergraphs.

- **Chapel HyperGraph Library (CHGL)** [3] that has been developed by the Pacific Northwest National Laboratory since 2018 and is released under MIT license. CHGL is a library for the emerging parallel language Chapel. The library provides the `AdjListHyperGraph` module that allows to store hypergraphs on shared and distributed memory. The library is not well documented and does not provide an easy mechanism for the 2-section and bipartite view analyses. However, it is worth mentioning for its functionality for parallel and distributed computing.
- **HyperX** [6] is a scalable framework for processing hypergraphs and learning algorithm built on top of Apache Spark. This library supports the same design model of GraphX, the Apache Spark API for graphs and graph-parallel computation written in Scala language. An interesting feature of this library is that it provides native support for hypergraph elaboration. The standard approach uses the bipartite or the 2-section representation of hypergraphs and exploits GraphX library, while HyperX directly processes hypergraph data obtaining significant speedup compared to the standard approach.
- **Pygraph** [11] is a pure Python library for graph manipulation released under the MIT license. It has almost all basic functionalities on graphs implemented but also supports hypergraphs by exposing the class `hypergraph`. This library does not provide any specific optimization and functionalities for hypergraphs.
- **Multihypergraph** [9] is a Python package for graphs released under GPL license. The library emphasizes the mathematical understanding of graphs rather than the algorithmic efficiency and provides support for hyper-edges, multi-edges, and looped-edges. This library provides only graph model memory definition and isomorphism functionalities without implementing any other functionalities and algorithms for graphs and hypergraphs.
- **HyperNetX** [5] is a Python preliminary library released in 2018 under the Battelle Memorial Institute licence[1]. The library generalizes traditional graph metrics (such as vertex and edge degrees, diameter, distance, etc.) to hypergraphs, and provides good documentation and tutorials. The library supports the bipartite representation of a hypergraph, along with the possibility to load hypergraphs from their bipartite view. Furthermore, it provides some simple visualization functionalities for hypergraph.
- **Halp** [2] is a Python software package providing both a directed and an undirected hypergraph implementation as well as several important and classical algorithms. The library is developed by Murali's Research Group at Virginia Tech released under GPL license. The library provides several statistics on hypergraphs and model transformations in graphs supported by the NetworkX Python library. In addition, several algorithms for hypergraphs, such as k-shortest-hyperpaths, random walk, directed paths, are implemented.
- **HyperGraphLib** [4] is a C++ implementation of hypergraphs that exploits the Boost Library, which also defines the library license. This library provides basic functionalities for hypergraphs and implements some simple metrics. Moreover, it also provides some isomorphisms functionalities and path-

---

[1] https://github.com/pnnl/HyperNetX/blob/master/LICENSE.rst.

finding algorithms. However, it does not implement any kind of hypergraph representations (such as bipartite or 2-section) nor software integration with other graph libraries.

- **Iper** [7] is a JavaScript library for hypergraphs released under MIT license. The library provides the definitions of hypergraphs and allows the user to define meta information for vertices. However, it does not include any kind of hypergraph transformation/representations and integration with other graph libraries for classical statistics and algorithms.
- **NetworkR** [10] is an R package with a set of functions for analyzing social and economic networks including hypergraphs. It includes analyses such as degree distribution, diameter, and density of the network, as well as microscopic level analysis such as power, influence, and centrality of individual nodes. The library does not provide support for meta information on vertices and hyperedges and provides only hypergraphs projection into graphs.
- **Gspbox** [1] is an easy to use Matlab toolbox that performs a wide variety of operations on a graph. It is based on spectral graph theory and many of the implemented features can scale to very large graphs. Gspbox supports hypergraphs modeling, including ability for hyperedges to have weights assigned, and for vertices to have coordinates in the space. The hypergraph manipulation is obtained by representing the model as a graph. For this reason, despite the fact that all graph functionalities are available, the library does not provide any kind of specific solutions or optimization for hypergraphs.

Overall, all the considered libraries settles a compromise between efficiency (which characterizes low level languages, such as C/C++) and easy-of-use/expressiveness (which characterized interpreted and/or scripting languages like Python and R).

In this work, we are proposing a library, which exploiting the Julia language ensures both efficiency and expressiveness. Julia is a new programming language developed at MIT [16]. The language uses a syntax similar to popular and easy-to-use scientific computing languages like Python or R. This means that experience in those languages can be directly applied in Julia by computational scientists [21,29]. Moreover, a distinguishing feature of Julia is that while keeping mathematics-oriented syntax it makes it possible to compile the code to a binary form. In result it means that the observed performance of Julia programs is very similar to that of C++, however with around 4 times less lines of code.

The library `SimpleHypergraphs.jl` is available on a GitHub public repository[2], where the library documentation is also provided[3]. Additionally, several tutorials are available in the form of Jupyter Notebooks[4]. This Section describes the library design and motivations behind its implementation. Furthermore, library functionalities of the 1.0 version will be discussed.

---

[2] https://github.com/pszufe/SimpleHypergraphs.jl.
[3] https://pszufe.github.io/SimpleHypergraphs.jl/latest/reference/.
[4] https://tinyurl.com/y5btobdk.

2.2 Definitions and Notation

Hypergraphs are natural generalization of well-known and widely used graphs. Formally, a hypergraph is an ordered pair $H = (V, E)$ where V is a set of vertices and E is a set of edges. Each edge is a non-empty subset of vertices; that is, $E \subseteq 2^V \setminus \{\emptyset\}$, where 2^V is the power set of V. We will use $n = |V|$ and $m = |E|$ for the size of the vertex set and, respectively, the edge set. Indeed, hypergraphs are generalization of graphs in which each edge is a two element subset of V; that is, hypergraph $G = (V, E)$ is a graph if $E \subseteq \binom{V}{2} \subseteq 2^V \setminus \{\emptyset\}$.

2.3 Library Design and Functionalities

SimpleHypergraphs.jl represents a hypergraph $H = (V, E)$ as an $n \times k$ matrix, where n is the number of vertices and k is the number of hyperedges. In other words, each row of the matrix is associated with a vertex and indicates the hyperedges the vertex belongs to. The proposed library stores in-memory a hypergraph using its matrix representation. Vertices and hyperedges are uniquely identified by progressive integer ids, corresponding to rows $(1, \ldots, n)$ and columns $(1, \ldots, k)$, respectively. Each position (i, j) of the matrix denotes the weight of vertex i within the hyperedge j. In addition, the library provides several constructors for defining meta information type and enables to attach meta-data values of arbitrary type to both vertices and hyperedges.

The library APIs are designed in similar fashion of the popular library for graph manipulation LightGraphs.jl, this provide to the programmers a familiar environment.

Hypergraph Constructors. Based on the previous consideration, the Julia hypergraph object is defined as:

$$\text{Hypergraph}\{T, \ V, \ E\} \ <: \ \text{AbstractMatrix}\{\text{Union}\{T, \ \text{Nothing}\}\}$$

where T represents the type of the weights stored in the structure while V and E are the type of values stored in the vertices and edges of the hypergraph, respectively.

Functions. SimpleHypergraphs.jl provides several accessing and manipulating functions:

- add_vertex!, adds a vertex to a given hypergraph H. Optionally, the vertex can be added to existing hyperedges. Additionally, a value can be stored with the vertex using the vertex_meta keyword parameter.
- set_vertex_meta!, sets a new meta value new_value for vertex id in H.
- get_vertex_meta, returns a meta value stored at vertex id in H.
- get_vertices, returns vertices from a H for a given hyperedge he_{id}.

The same functionalities are provided for the hyperedges.

Hypergraph Transformations. The library provides two hypergraph transformations into the corresponding graph representation:

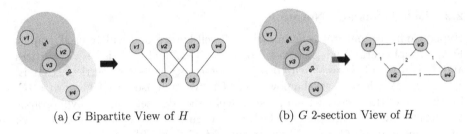

(a) G Bipartite View of H (b) G 2-section View of H

Fig. 1. (H)ypergraph transformations.

1. `BipartiteView` is a bipartite representation of a hypergraph H. As described in Bretto [17], this representation is an incidence graph of hypergraph $H = (V, E)$; that is, a bipartite graph $IG(H)$ with vertex set $S = V \cup E$, and where $v \in V$ and $e \in E$ are adjacent if and only if $v \in e$. Figure 1a (on the left) depicts a simple example of bipartite view.

2. `TwoSectionView` is a 2-section representation of a hypergraph H. As described in Bretto [17], this representation of a hypergraph $H = (V, E)$, denoted by $[H]_2$, is a graph whose vertices are the vertices of H and where two distinct vertices form an edge if and only if they are in the same hyperedge of H. As a result, each hyperedge from H occurs as a complete graph in G. The weight of an edge corresponds to the number of hyperedges that contain both the endpoints of the edge.

   Figure 1b (on the right) shows a simple example of the 2-section view.

Both `Views` are instances of the `AbstractGraph` graph object defined by the `LightGraphs.jl` library [8]. When the view is materialized—according to `LightGraphs.jl` specifics—the generated graph does not include any meta information.

Hypergraph I/O. The library currently offers a basic mechanism to load/save a hypergraph from/to a stream. Given hypergraph H is stored using the following format. The first line consists of n and k, the number of vertices and, respectively, the number of edges of H. The following k rows describe the actual structure of H. Each row represents one hyperedge as a list of all vertex-weight pairs within that hyperedge.

2.4 Hypergraph Modularity

One of the most important properties of complex networks is their community structure, that is, the organization of vertices in clusters, with many edges joining vertices of the same cluster and comparatively few edges joining vertices of different clusters. In social networks communities may represent groups by interest, in citation networks they correspond to related papers, in the Web communities are formed by pages on related topics, etc. Being able to identify communities in a network could help us to exploit this network more effectively. In our example, clusters in *Yelp* hypergraph may help to find similar restaurants, discovering users with similar interests that is important for targeted advertisement.

The key ingredient for many clustering algorithms is modularity, which is at the same time a global criterion to define communities, a quality function of community detection algorithms, and a way to measure the presence of community structure in a network. Modularity was introduced by Newman and Girvan [27] and it is based on the comparison between the actual density of edges inside a community and the density one would expect to have if the vertices of the graph were attached at random, regardless of community structure. The modularity function was recently generalized to hypergraphs [24] but no fast, heuristic algorithms are developed yet for this hypergraph counterpart. Our goal is to propose a number of potential solutions in the forthcoming paper and in this paper we present applicability of this method that has been already implemented in the `SimpleHypergraphs.jl` library.

3   Use Case—*Yelp* Dataset

In this section, we present a practical application of the `SimpleHypergraphs.jl` library. We especially focus on and analyze *Yelp* dataset consisting of reviews of restaurants. A natural representation of such data is a hypergraph in which vertices are associated with restaurants and hyperedges are associated with reviewers who reviewed various restaurants. The topology of this hypergraph allows us to find clusters of restaurants that are commonly reviewed together. As hypergraph clustering is an example of an unsupervised learning technique, our goal is to learn if such clusters are related to some natural characteristics of underlying restaurants. Such analysis allows us to better understand which factors (ground-truth) influence the changes that two restaurants are reviewed together. To that end we propose a methodology to measure and then to compare the results of hypergraph clustering against various possible ground-truth variables (here the main challenge is to develop a measure comparable across different ground truths). Since the *Yelp* dataset is used only as an example, the proposed approach can be used to identify ground-truths in other datasets that are represented as a hypergraph. As side effect of this use case, we also show that the hypegraph based approach conveys more information about the ground-truth properties of a hypergraph than a standard graph analysis approach. In particular, we compare the results obtained for hypergraphs with the corresponding results for 2-section, and show that hypergraph clusters provide uniformly more information than their graph counterpart. Additionally, when analyzing the data we consider different sub-hypegraphs, namely, we examine hypergraphs containing only reports with a given number of stars, from 1 to 5. This approach sheds some light on how review linkages are formed; in particular, we test how the mechanism behind those linkages differs across different review classes.

An interesting property that is worth to investigate, typical to many such networks, is the community structure, that is, the division of networks into groups of vertices that are similar among themselves but dissimilar from the rest of the network. The capability to detect the partitioning of a network into communities can give important insights into the organization and behaviour of the system that the network models.

Table 1. *Yelp* entities contained in the dataset.

Data	Instances	Description
Business	192,609	Business data including location, attributes, and categories
User	1,637,138	User data including the user's friend mapping and all the metadata associated with the user
Review	6,685,900	Full review text including the user_id that wrote the review and the business_id the review is written for
Picture	200,000	Photo data including caption and classification (one of "food", "drink", "menu", "inside" or "outside")
Tip	1,223,094	Tips written by users on businesses. Tips are shorter than reviews and tend to convey quick suggestions
Check-in	192,609	Aggregated check-ins over time for each business

3.1 The Yelp Open Dataset

Yelp is an online platform where customers can share their experiences about local businesses by posting reviews, tips, photos, and videos. It allows businesses and customers to engage and transact [12]. Every year, the Yelp Inc. Company releases part of their data as an open dataset to grant the scientific community to conduct research and analysis on them. Some interesting articles that use the Yelp dataset for their analysis can be found in [22,23,25,26]. As a use case, we analyzed the 2019 Yelp Challenge dataset [13], containing information about businesses, reviews, and users. Table 1 describes all the accessible dataset *entities*. A more detailed description can be found on the official page [14].

Figure 3 (on the left) presents business categories distribution, where a category is a label describing the typology of the business such as *Bars* or *Shopping* along with the number of reviews associated with each category. It highlights the category distribution evaluated over all businesses. As clearly visible from the plot, the most common business typology is *Restaurant*. For this reason, we focused our analysis on this business subgroup. Figure 3 (on the right) shows the category distribution evaluated only within the *Restaurant* macro-category. Both Figures show top-20 most common categories.

3.2 The Yelp Hypergraph

We model *Yelp* dataset using a hypergraph $H = (V, E)$, where V represents businesses and E represents users of Yelp. In particular, each hyperedge representing user u contains businesses u has written at least one review for. Figure 2 shows an example hypergraph representing a Yelp data subset. As shown in the figure, the hypergraph H is defined by four businesses ($V = \{b_1, b_2, b_3, b_4\}$) and three users ($E = \{u_1, u_2, u_3\}$). For instance, hyperedge u_1 connects businesses b_1, b_2, and b_4, as the corresponding user have written reviews for each of the listed business.

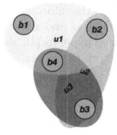

Fig. 2. Yelp Hypergraph defined by the users reviews.

Since processing the entire Yelp dataset is a heavy computationally task, for our purpose we have decided to explore only a subset of it. We have modelled the Yelp hypergraph according to the following building strategies:

1. **yelpdataset1** is a random selection of reviews of specific sizes. It is worth mentioning that such selection of reviews defines also the number of businesses involved. Indeed, our analysis are executed on connected hypergraphs that are obtained by removing isolated vertices and small components.
2. **yelpdataset2** is a subset of those businesses that belong to the category "restaurant" (note that some businesses have more than one category; in such cases we select one category from its categories set according to the frequencies (highest) in the whole dataset).

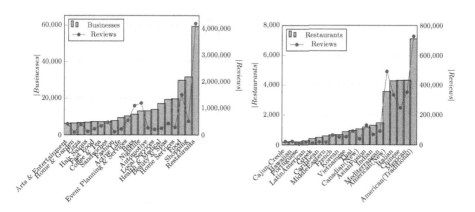

Fig. 3. Businesses (left) and Restaurants (right) distribution and number of reviews associated with each category.

3.3   Results

We are interested in the following two research questions. First of all, our goal was to investigate whether modelling the *Yelp* dataset with hypergraphs gives qualitatively more information than looking at the corresponding 2-section graph representation. Then we compared the information provided by the three hyper-graphs consisting of positive, neutral and negative reviews. In this case, the research question is: are the three hypergraphs similar or different? In order to answer the two questions, we set up two experiments explained below.

Experiment I: "Forecasting stars". This experiment tries to forecast the number of stars of a given business v, based on the information available in the local neighbourhood of v. Two different strategies have been developed, one is based on the information provided by hypergraph H defined above, and one is based on the information provided by the weighted 2-section of the same hypergraph. Here, the weight of an edge (u, v) corresponds to the number of users that reviewed both u and v, that is, the number of hyperedges that contain both u and v.

For the first strategy (on hypergraph H), for each business u, we first compute the average number of stars for all hyperedges containing u; in each hyperedge e, the average is computed excluding u. This corresponds to the typical rating given by the user associated with e. Then, the forecast for the number of stars of u is obtained as the average over the values computed at the previous step. In other words, the forecast of the number of stars of u is the average over the averages in each hyperedge involving u. Formally,

$$s'_i(u) = \frac{1}{|E(u)|} \sum_{e \in E(u)} \left(\frac{1}{|e| - 1} \sum_{v \in e, v \neq u} s(v) \right),$$

where $s(v)$ denotes the number of stars associated to v, $E(v)$ denotes the set of hyperedges that contains v, and $s'_i(u)$ denotes the forecasted value for u for strategy i.

The second strategy exploits the weighted 2-section graph. In this case, the forecast of the number of stars of u is the weighted average over the neighborhood of u. Formally,

$$s'_2(u) = \frac{\sum\limits_{e=(u,v) \in E} s(v)w(e)}{\sum\limits_{e=(u,v) \in E} w(e)},$$

where $w(e)$ denotes the weight of edge e.

In order to compare the two strategies, we computed their average error as follows:

$$err_i = \frac{\sum\limits_{u \in V} |s(u) - s'_i(u)|}{|V|}.$$

We performed our experiment on several instances of **yelpdataset1**, varying the number of reviews used. The left side of Fig. 4 depicts the obtained results for stars' forecast experiment. The error value err_2 using the weighted 2-section graph is always greater than the error value err_1 obtained for the hypergraph representation.

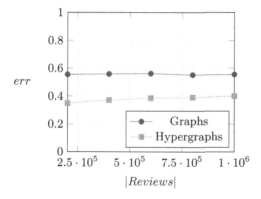

Fig. 4. Stars' forecast varying the dimension of the reviews set on **yelpdataset1**.

We also experimented with forecasting on **yelpdataset2**, obtaining similar results; the error for graphs is always close to 0.6 while the error for hypergraphs is always close to 0.5. Both the results are promising since the average number of stars obtained by businesses are around 0.5 and so it is important to be able to accurately predict low rated instances. Our experiment shows that the information provided by the hypergraph is more accurate than the information provided by the corresponding weighted 2-section.

Experiment II: Positive, Neutral, and Negative Reviews. The second experiment examines the amount of information given by different kind of reviews, depending on the number of stars associated to them. We used **yelpdataset2** but due to the performance issues, we restricted the set of businesses to restaurants category, as described in Sect. 3.2. Five hypergraphs were built after partitioning the reviews into five categories: 1 stars, 2 stars, . . . , 5 stars.

In the dataset we have 342,044 1-star reviews, 281,307 2-star reviews, 402,053 3-star reviews, 791,068 4-star reviews and 1,188,558 5-star reviews. Hence, we decided to build five hypergraphs, one for each set of reviews. Henceforth, for $i = 1, 2, \ldots, 5$, we will denote by H_i, the hypergraph generated using the set of reviews having i stars and by G_i the corresponding 2-section view graph.

Table 2. Graphs statistics.

Stars	H_i ($|V|$; $|E|$)	G_i ($|V|$; $|E|$)	G_i modularity	G_i triangles
1	(29479; 244671)	(29479; 240412)	0.6210	1,158,341
2	(28055; 173140)	(28055; 484527)	0.7173	6,491,497
3	(30369; 177792)	(30369; 2636712)	0.6616	289,584,451
4	(32987; 301578)	(32987; 4384044)	0.6857	404,709,664
5	(32558; 590320)	(32558; 2187473)	0.6657	104,128,714

Table 3. Hypergraph modularities for various number of stars and various ground truth based partitioning conditioned on properties of restaurants.

Stars	H_i ($|V|$; $|E|$)	City	State	Alcohol	Noise level	Take Out	Category
1	(29479; 244671)	0.8833	0.9562	0.8166	0.8104	0.8176	0.8163
2	(28055; 173140)	0.8582	0.9462	0.7744	0.7651	0.7731	0.7702
3	(30369; 177792)	0.8132	0.9226	0.7075	0.6940	0.6966	0.6965
4	(32987; 301578)	0.7812	0.9081	0.6573	0.6385	0.6419	0.6400
5	(32558; 590320)	0.8027	0.9145	0.6963	0.6797	0.6894	0.6841
ALL	(35856; 950488)	0.7500	0.8985	0.6162	0.5919	0.6013	0.5967

First, we computed some statistics on the five hypergraphs and their corresponding 2-section views. The collected information can be found in Table 2. This preliminary analysis shows that the five hypergraphs/graphs are quite different. For instance, for the 2-section graphs, the number of edges, and the number of triangles exhibit a "bell-shaped" trend as a function of the number of stars. As a result, we shift our attention to their ability to detect the community structure, that is, the division of the vertex set into groups of restaurants that are similar among themselves but dissimilar from the rest of the network. In order to evaluate this feature, we decided to run some community detection algorithms on each graph/hypergraph. We then compared the obtained results with a ground truth restaurant partitioning, based on the "type of cuisine" provided by the system. This ground truth partitioning is composed of 55 categories of which the largest (American Traditional) comprises 7,107 restaurants.

The Table 3 contains modularity values for various partitionings of the hypergraph. In order to calculate modularities we used approach presented in [24] that we have implemented as the `modularity` function in the SimpleHypergraphs.jl library. One can see that the modularity is strongest when we uses city or state to partition the hypergraph. This means that people doing reviews usually use restaurants within the same city and if restaurants in different cities are reviewed by a single person they are usually in the same state. It can be noted that reviews with one star have the strongest modularity values across all partitionings. This probably means that there is a group of people who have a stronger tendency to submit negative scores on the base of some ground-truth property of a restaurant.

Several community detection algorithms have been proposed in the literature. A review of the various methods available can be found, for example, in [15,20]. For graphs, we decided to opt for a label propagation (LP) strategy proposed by Raghavan et al. [28]. This strategy can be summarized as follows: each node is initially given a unique label (initialization); at each iteration, each node is updated by choosing the label which is the most frequent among its neighbours (propagation rule)—if multiple choices are possible (as, for example, at the very beginning), one among the candidate labels is picked randomly. The algorithm terminates at the first iteration that leaves the label configuration unchanged or after the predefined number of iterations (termination criteria). We exploited the LP implementation provided by the Julia LightGraphs library [8].

For hypergraphs, we implemented an ad-hoc label propagation strategy which generalizes the algorithm in [28] for hypergraphs. The proposed algorithm shares the initialization phase as well as the termination criteria with the standard label propagation algorithm. On the other hand the propagation rule is, in this case, composed of two phases: hyperedge labelling and vertex labelling. During the hyperedge labeling phase, labels of hyperedges are updated according to the most frequent label among the vertices that belong to the edge. Then, during the vertex labeling phase, label of each vertex is updated by choosing the label that is the most frequent among the hyperedges it belongs to.

Both algorithms have been executed setting the maximum number of iterations to 100. We compared the partitions obtained running the label propagation strategies described above with the ground truth partition in order to learn how much they are related. Several measures to evaluate the correlation between the two partitions have been borrowed from information theory. In particular, by considering a partition as a probability distribution, the *Normalized Mutual Information* (NMI) is often used to measure their correlation. Several variants of the NMI have been defined (see, for example, [30] for a detailed discussion). In this paper we use the *sum* variant which is defined as follows:

$$NMI(X,Y) = \frac{I(X,Y)}{H(X) + H(Y)}, \tag{1}$$

where $I(X,Y)$ denotes the *Mutual Information* (that is, the shared information between the two distributions X and Y) and $H(X)$ denotes the Shannon Entropy (that is, the information contained in the distribution) of X. NMI enjoys several interesting properties: namely it is a *metric* and lies within a fixed range $[0, 1]$. Specifically it equals 1 if the partitions are identical whereas it has an expected value of 0 if the two partitions are independent.

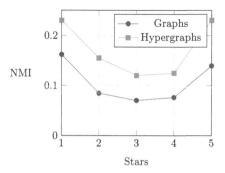

Fig. 5. Then NMI between the ground truth partition and the 10 partitions obtained running the label propagation algorithm on the five hypergraphs and on the corresponding 2-section views.

Results appear in Fig. 5. Although the correlation in general is not very high (the best result is 0.23 for H_5), the figure provides two interesting points. First, in all the five considered cases, the quality of partitioning provided by hypergraphs is always better than that provided by the corresponding 2-section view graph. Moreover, also in this case, the results appear in the form of an "inverted bell shape" (the best results in this case are given by the two external values). In a sense, very good as well as very bad reviews are much better able to identify restaurants genre.

4   Conclusion

In this work we have presented a novel library for the manipulation and analysis of hypergraph structures. Hypergraphs have been shown to be much better than standard graphs to model many natural phenomena, such as collaborative activities, which involves group based interactions.

The library, named `SimpleHypergraphs.jl`, provides Hypergraph views built exploiting the popular package `LightGraphs.jl` a Julia library for graphs manipulation. Several functionalities for the I/O, the manipulation and the transformation of hypergraphs have already been developed and are available on a public GitHub repository. In addition, the library enables the user defining meta information type as well as attaching meta-data values of arbitrary type to both vertices and hyperedges. This approach enables for an efficient analysis of structural properties of the network, combined to the possibility to perform semantic analysis based on the attached meta-data. The Yelp dataset case studies show that it scales well when analyzing thousands of nodes connected by millions of edges. We plan to expand the library by developing novel functionalities and a visualization engine which will enable the exploration of the hypergraph networks as well as of the enclosed meta-information. We have presented also a case study based on the Yelp dataset showing some of the functionalities available on `SimpleHypergraphs.jl` and, at the same time, that hypergraph networks convey much information with respect to their corresponding graph representation.

References

1. GSPBox, MATLAB (2019). https://github.com/epfl-lts2/gspbox
2. HALP, Python (2019). https://github.com/Murali-group/halp
3. HyperGaph, Chapel (2019). https://github.com/pnnl/chgl (2019)
4. HyperGraphLib, C++ (2019). https://github.com/alex-87/HyperGraphLib
5. HyperNetX, Python (2019). https://github.com/pnnl/HyperNetX
6. HyperX, Scala (2019). https://github.com/jinhuang/hyperx
7. IPER, JavaScript (2019). https://github.com/fibo/iper
8. LightGraphs.jl, Julia (2019). https://github.com/JuliaGraphs/LightGraphs.jl
9. Multihypergraph, Python (2019). https://github.com/vaibhavkarve/multihypergraph
10. networkR, R (2019). https://github.com/O1sims/networkR
11. PyGraph, Python (2019). https://github.com/jciskey/pygraph
12. Yelp (2019). https://www.reuters.com/finance/stocks/company-profile/YELP.N
13. Yelp-dataset (2019). https://www.yelp.com/dataset/challenge

14. Yelp-dataset-docs (2019). https://www.yelp.com/dataset/documentation/main

15. Antelmi, A., Cordasco, G., Spagnuolo, C., Vicidomini, L.: On evaluating graph partitioning algorithms for distributed agent based models on networks. In: Hunold, S., et al. (eds.) Euro-Par 2015. LNCS, vol. 9523, pp. 367–378. Springer, Cham (2015). https://doi.org/10.1007/978-3-319-27308-2_30

16. Bezanson, J., Edelman, A., Karpinski, S., Shah, V.B.: Julia: a fresh approach to numerical computing. SIAM Rev. **59**(1), 65–98 (2017)

17. Bretto, A.: Hypergraph Theory: An Introduction. Springer, Cham (2013). https://doi.org/10.1007/978-3-319-00080-0

18. Cordasco, G., Spagnuolo, C., Scarano, V.: Toward the new version of D-MASON: efficiency, effectiveness and correctness in parallel and distributed agent-based simulations. In: 2016 IEEE International Parallel and Distributed Processing Symposium Workshops (IPDPSW), pp. 1803–1812 (2016)

19. Cordasco, G., De Chiara, R., Raia, F., Scarano, V., Spagnuolo, C., Vicidomini, L.: Designing computational steering facilities for distributed agent based simulations. In: Proceedings of the 1st ACM SIGSIM Conference on Principles of Advanced Discrete Simulation, pp. 385–390 (2013)

20. Danon, L., Díaz-guilera, A., Duch, J.: Comparing community structure identification. J. Stat. Mech. Theory Exp. (2005)

21. Edelman, A.: Julia: a fresh approach to technical computing and data processing. Technical report, Massachusetts Institute of Technology, Cambridge (2019)

22. Gulati, A., Eirinaki, M.: Influence propagation for social graph-based recommendations. In: 2018 IEEE International Conference on Big Data (Big Data), pp. 2180–2189 (2018)

23. Ji, Z., Pi, H., Wei, W., Xiong, B., Woźniak, M., Damasevicius, R.: Recommendation based on review texts and social communities: a hybrid model. IEEE Access **7**, 40416–40427 (2019)

24. Kaminski, B., Poulin, V., Pralat, P., Szufel, P., Theberge, F.: Clustering via hypergraph modularity. arXiv preprint arXiv:1810.04816 (2018)

25. Li, R., Jiang, J.Y., Ju, C.J.T., Wang, W.: CORALS: who are my potential new customers? Tapping into the wisdom of customers' decisions. In: Proceedings of the Twelfth ACM International Conference on Web Search and Data Mining, WSDM 2019, pp. 69–77 (2019)

26. Lu, X., Qu, J., Jiang, Y., Zhao, Y.: Should i invest it?: predicting future success of yelp restaurants. In: Proceedings of the Practice and Experience on Advanced Research Computing, PEARC 2018, pp. 64:1–64:6 (2018)

27. Newman, M.E., Girvan, M.: Finding and evaluating community structure in networks. Phys. Rev. E **69**(2), 026113 (2004)

28. Raghavan, U.N., Albert, R., Kumara, S.: Near linear time algorithm to detect community structures in large-scale networks. Phys. Rev. E Stat. Nonlinear Soft Matter Phys. **76** (2007)

29. Regier, J., et al.: Cataloging the visible universe through Bayesian inference in Julia at Petascale. J. Parallel Distrib. Comput. (2019)

30. Vinh, N.X., Epps, J., Bailey, J.: Information theoretic measures for clusterings comparison: variants, properties, normalization and correction for chance. J. Mach. Learn. Res. **11**, 2837–2854 (2010)

Author Index

Printed in the United States
By Bookmasters